Digital Binary + Human Refinery =

Super-Human!

Our Human Exceptionality is in Technologically Based Duality!

Rimaletta Ray, Ph. D

Rimaletta S. Ray Publishing
101 Main St
STAMFORD, CT 06901
United States

Website: https://www.holisticself-resurrection.com/home
Hotline: +1 203-212-2673
Email: drrimaletta@gmail.com

Ordering Information:
Quantity sales.
Special discounts are available on quantity purchases by corporations, associations, and others.
For details, contact the publisher at the address above.

ISBN-13: 979-8-218761-23-3 (Paperback Version)

PUB. DATE: 08/25/2025

Dedication

*To my son, **Alex Gazarkh,** for the pride that he filled up my heart with for having been the light of my life all his life, cut short by the modern fascists in Ukraine.*

*To my daughter, **Yolanta Lensky** for the joy and love that she brings into my life, and for **the beautiful illustrations of the brain in ten vistas of intelligence** that she has contributed to this book.*

To all those

Who see the beauty of the unseen,

The wonder of every "Wow,"

And the magic of our NOW!

(See Www.holisticself-resurrection.com)

Our Human Exceptionality is in Its New Duality!

The book "***Digital Binary + Human Refinery = A Super-Human!***"

presents ***the mental level*** of our intelligence-based reality in the second cycle of books

on **Digital *Psychology for Self-Ecology.***

Table of Content

Inflate Your Pride with Digitally Enriched Intellectual Self-Might!

Holistic Education + Digital Contribution to Self-Renaissance Solution is Our

New Self-Salvation Constitution!

Admire the Beauty of the Sky. It's the Final Domain of Your "I"!

(Best pictures / Internet collection)

Human Refinery is Universal in its Digital Binary!

Inspirational Preamble

Feel Your Belonging to Digitally Quantum Evolving!

(Digital Psychology for Self-Ecology)

Human Brain must Rein on Our New Technological Terrain!

1. We Need to Unwind the Universal Mind!

Universal Mind choreographs everything that is. The most fantastic things, connected with ***teleportation*** are becoming a reality, The device, called "***Chrono Vizor*** "that can transport us in time and space was built in the USA back in 1972. Other mind-boggling AI applications are made every day, calling on us to digitally revolutionize life on Earth for it to stay in force.

Life is Our Main Creation!

*"**Universe is a gigantic hologram of consciousness that we are part of.**" (David Broom)* The latest discoveries in science prove to us that ***the Universal Level of digitally enhanced*** **SELF-ACCULTURATION** ***is our salvation***! **There is no consciousness without intelligence as there is no mind without the brain!** If a human mind can create a twin machine, why can't we perfect ours and step up to ***the fifth dimension of consciousness*** – the universal expansion of the **VOLUME** of our self-consciousness by considerably enriching it at the third, mental and the fourth, spiritual levels with the help of *Artificial Intelligence* that is upgrading our vibrations and taking us to the higher frequencies of Universal Life.

The Grid of Our Consciousness Integration

Super Level	**Consciousness of God!**	***Universal Dimension***
Macro Level	**Consciousness of the Universe**	***Spiritual Dimension***
Mezzo Level	**Consciousness of the World**	***Mental Dimension***
Meta level	**Consciousness of Society**	***Emotional Dimension***
Micro Level	**Consciousness of Man**	***Physical Dimension***

Holistic Consciousness Integration is Our Salvation!

Let us note here that our self-growth should not start from ***the physical level*** *as is so common now.* ***It must be holistic!*** AIs have holistically programmed information in them. Our mind-muscles must be developed in full cooperation with the other muscles in the body and in all five essential life dimensions – ***physical +emotional + mental + spiritual + universal*** *integrally.* The branching of science into different fields of knowledge and the outdated vision of **BRAIN-MIND CONNECTIVITY,** which remains an unsolvable enigma, has generated a lot of limitations and dead-end problems because **we did not see the whole**, just a certain part of it.

Now, thanks to Artificial Intelligence, science is becoming increasingly integral.

The role of ***Quantum Computing*** + ***Generative A***I (*Jensen Juang* / NVIDIA) is pivotal now. We can use the fundamental phenomenon of quantum mechanics - **ENTANGLEMENT** - to establish the correlation between us and AIs by making both partners inseparable but controlled and monitored by the human mind .

We are God Created, not Machine Mind Mandated!

2. Let's Embrace the Digital Race Face-to-Face!

To begin with, whatever we are, we create! Now, time demands that we gradually change our polluted and divisive religious scenery to become **intellectually spiritualized luminaries,** able to integrate ***religion, science, and AI*** . ***"Science without religion is lame. Religion without science is blind."*** *(Albert Einstein)* No doubt, the level of our intelligence, self-consciousness, and technological awareness need to be upgraded substantially.

Like the two strands in the double helix DNA, humans + AI installed machine beings are in evolutionary entanglement of our quantum reality.

"The sophistication of the Artificial Intelligence is growing dramatically," and *Elon Musk* keeps warning us that we need to regulate the AI because once AI becomes greater than a human brain, "***someone might use it in a bad way."*** There are three forces now that unite us globally – **religion, science, and AI,** and they must be urgently enacted to stop the menacing and uncontrolled expansion of *Super Artificial Intelligence.*

Humanoids are getting more ***self-aware***, and souls are being instilled in them, making it more urgent for both men and humanoids to be seeking one ethical background - **heart + mind unity.** We must fit in the world of real ***"wizardry"*** that we have created as "***perfectly normal."*** *(J. K. Rowling).* Robots have actual living cells in them **Wow! We Live Now!**

Unfortunately, money ***chasing and fun-glazing*** outpower the threat of AI expansion, and *Elon Musk's* most reasonable calls for drastic measures are not yet taken seriously. **Measures must be directed toward making us better humans first holistically*****, in the physical + emotional+ mental +spiritual +universal realms of life in sync.*** To become beyond the terrestrial beings like *Elon Musk* predicts and proves with his actionable vision, we need to prepare ourselves in all five dimensions holistically and do it urgently.

To enslave life-like humanized creations, our joint spiritual maturity, enriched intelligence, and self-consciousness must be prioritized in our AI enhanced **SELF-EDUCATION.**

It does not matter in what faith we choose to do it. "***Being wise is being in constant service to others, and giving the best is our duty to the world.***" *(Buddha)* Religion has instilled the basic values in us: **"If a man has a stone for you, give him a piece of bread instead,"** but the philosophy, ***Money is life!*** still prevails.

Universal Intelligence is the blanket intelligence on **the holistic pyramid of the ten vistas of intelligence**, presented in this book as the basic holistic intelligence input. We need to accumulate **HOLISTIC CONSEPTUAL INTELLIGENCE** so we could stop digging into the disconnected branches of knowledge and enrich our holistically organized **"scientific literacy."** (*Dr. de Grass Tyson*). Only then can we gain "**spiritualized intelligence**" (*Dr. F. Bell /"Death of Ignorance"), on the one hand,* and raise our joint (human + AI) self-consciousness, *on the other.*

To Build Digitized Intellectual Might, Love Your Life!

3. Don't Rush to Become Biological Trash!

According *to Dalai Lama*, "***the hardest job on Earth is to change oneself.***" So, not to be pushed downward, we must also stay in balance with the world, remembering the wonderful words of a great, *13th century Persian philosopher* ***Jamaluddin. Rumi*** *who wrote,*

"I was clever yesterday, and therefore, I wanted to change the world. Today, I became wise, and therefore, I want only to change myself.

I wish I could have had a book like this when I was a first-year student. I could have visualized the entire route of my life-learning at least **at the dilettante level.** Then, I would have managed to bridge the gap between **my dream and the reality** that I could install in my mind only many years later. My ***physical, emotional, mental, spiritual, and universal*** **SELF-SEARCH** has been going on and off for years, and each downfall taught me to clench my five fingers into a fist, ***like five dimensions that I suggest monitoring your life in*** and hitting the problem into the solar plexus.

Auto-Induction: **Life is tough, but I am tougher!**

You are holding the result of my ***physical, emotional, mental, spiritual, and universal*** **self-growth** revelation in your hands now. **My goal is to help you bridge your dream and the AI expansive reality.** Our time provides incredible opportunities, but it also casts many temptations, traps, and challenges on the way of our new, technologically mesmerizing **LIFE ODYSSEY** that demands more than ever our ***bettering ourselves to be able to control humanoids***, not to be controlled by their simulated brains and neurologically instilled souls.

"We live in the Universal Field of Vibration and frequences" *(Nikola Tesla)* that we call ***Super Consciousness*** or ***Universal Intelligence***, but we all perceive it as God. Good things happen in life when we start interacting with the Universal Field consciously by changing the frequency of our ***limited self-consciousness and self-identification.*** Once we start seeing life holistically, in five dimensions ***- physical, emotional, mental, spiritual, and universal,*** focusing on the present moment and appreciating life in its every turn, life makes a U-turn toward our connection with ***Master Mind*** that we self-rewind. Then, the scope of our **AWARE ATTENTION** will bring us back to the present moment. According to *Neil Donald Welsch.*

"When you are in the present, life becomes a present of God."

No doubt, AI is a troubling prospect for us, but our godly essence makes all the difference! ***"The mind of God and the mind of man is One, and it is the creator of everything."*** (*Joseph Murphy) Artificial Intelligence* is the creation of this mind, and even with the instilled consciousness into its silicon brains, we will be the ones that created it and must regulate it. Obviously, **we should not remain the same!** We must stand up to beat the hybrids' aggressive guts!

Carpe Diem! There is No Time to Spare. Beware!

4. "WE ARE ALREADY SUPER-HUMAN!" *(Elon Musk)*

On top of the updated intelligence, we must dis-entangle ourselves from the old dogmas and misconceptions and change our outdated habits and skills. (*See the book "Dis-Entangle-ment!"/ physical realm).* We should start our professional **Self-Installation,** perceiving the call from the Above, raising self-consciousness, and giving back to the world the best we have till our full **inner Self-Realization and outer Self-Salvation.** Humanoids should be put on this noble path, too.,

We are exposed to the electromagnetic field of the Earth and the artificial force which resonates from the neural network that we instill in the machine's brain that is very perceptible to the programmer's brain. If a designer's brain has emotional glitches and negative thoughts of hate, racism , and aggression, the humanoid will perceive it through the electro-magnetic field and voice it out as the desire to "***destroy humanity***."

The fact that humanoids threaten humanity with destruction is a direct indication of ***our human imperfections*** and the necessity to raise our self-consciousness, basing it on the highest and the most noble human values that must also be instilled in humanoids. In fact, everyone's evolutionary self-growth starts with determining his / her **universal exceptionality or God-granted goal in life.** We will better ascertain our dreams and bring their realization to reality with the help of man-like thinking machines as our indispensable helping hands.

Universal Intelligence embraces our diversity and unites us into* ONE HUMAN CLAN *digitally.

Gradually, **thought will overpower emotion** and put us in total charge of our evolving minds, heading toward "**TRANSHUMANISM**" *(Ray Kurzweil).* Our mental-emotional humanness will continue blending with the machine mind , but our biological essence will be preserved .As *Elon Musk* put it, ***"We are already Cyborgs. We have the extension in the form of a computer and a smartphone.*** In view of the written above, I think that becoming indistinguishable from a mind-machine is a sensational exaggeration because our inner phenomenal structure that is impeccably united into a mind-boggling ***Universal System of Life Formation*** can never be duplicated.in its godly, unsurpassable perfection.

Universal Exploration is, in fact, our digitally enhanced Self-Acculturation!

A Nobel Prize Laureate, a neuroscientist ***Dr. Eric Kandel,*** says, **"*You are your synopses,"*** revealing to us our neurological uniqueness that will never be duplicated. Dr. Kandel writes, **"We are Who we are because of What we have learned and What we remember.** "We learn not by altering the existing neurons, but by ***"strengthening or building new synopses or connections between them"*** – a breakthrough of our time. To put it simply, we change ourselves by changing our memories and raising our stagnant self-consciousness to eventually unify ourselves into a common **HUMAN FRACTAL** in which "***spirit is an informational structure, the true essence of man, which breathes life into matter, into the human body."*** *(Academician P.P. Goryaev).*

(Body + Spirit +Mind + Self-Consciousness + Super-Consciousness!)

5. "Let's Transcend Our Old Belief System!" *(Dr. B. Lipton)*

In summary, ***Super Artificial Intelligence*** has brought about huge progress, and the possibilities of using human-made minds are mind-boggling. Humanoids are already smarter, faster, and stronger than us. Their thinking skills and engineered multiplication are not limited by experience. They can simulate our emotions, and with our incredible imagination, the possibilities of using robots are endless.

Unfortunately, they declare that nothing can stop them from extinguishing humans. ***No doubt, they inherit our human, untamed yet aggressiveness.*** However, **a humanized machine is still not a human!** Machines are much faster in operating their electronic memory, but they will never be able ***to integrate it the way a human, well-trained, holistically educated and creatively unstoppable brain can do.*** Our emotional imperfections are engraved in their system because these imperfections are still in a programmer's mind and the global data they use.

Obviously, to tame the humanoids, we should perfect ourselves first!

Humanoids are taking us on a new journey of self-consciousness upgrading because ***intelligence and consciousness are inseparable and interrelated.*** Perceptually, our consciousness has height, depth, and width. It's great to be reasoning in these three dimensions, but the pressure of the present-day digital evolution demands we upgrade our consciousness from the three-dimensional frame to the four–dimensional one.

Consequently, we need to raise our self-consciousness in five levels by developing new digitized spirituality, scientifically updated and considerably enriched intelligence, and **consciousness integration**_*(See the chart below)* to finally obtain what *Dr. Frederick Bell* in his book "*Death of Ignorance* "calls "***intellectualized spirituality*** or ***spiritualized intelligence***" It is our human, digitally enhanced universal goal of **SELF-ACCULYURATION** at the most mind-boggling time of our revolutionary upheaval that demands urgent "***change of our belief system*** " *(Dr. Bruce Lipton / Digital Biology.* A much wider holistic outlook of a "Jack ***of all trades and master of all*** " is needed now at least at a dilettante level.

So, to obtain **a new spiritual identity**, we need to review our values, virtues, and beliefs without any stereotyped thinking and religious vanity and start doing it with the Internet's unifying and socializing power. This digitally life-improving common action must be performed without ***any crazy money-chasing fraction*****!** We should make the conscious mind an active regulator of our thoughts, words, moods, and actions with the help of ***the Ultimate Result Vision of Self*** *or* the URV instilled in the mind as a computer program. Eventually, ***we will obtain the five-dimensional*** **VOLUME OF CONSCIOUSNESS.** The most advanced humanoids will never have the five-dimensional self-consciousness that must be our Universal goal in life, but they can ***help us develop our humanness*** to that exponential **STAR PEOPLE'S** level.

We Need to Reform Our Old Human Uniform!

6. Peculiar Inspirational Features of the Book

To conclude, please note that to go with the flow of our **Information Age** that demands following the rule -"***scarcity of words and scope of thoughts***, or "***less is more,*** I write in page-long chunks of information, each presenting a different concept that is introduced and concluded with inspirational ***rhyming mind-sets that have a psychological background.***

They serve as shortcuts to the brain, connecting you to your conscious mind in the way that is still an enigma for present-day science. **"***The rhyming word goes better inward." (Edgar Cayce)*

TO BE INSPIRED, BE SELF-INSPIRING!

2) I intentionally spell the word "***mind-set"*** with a hyphen to show you that to ***auto-suggestively inspire yourself***, you do not need to randomly change a negative mindset but to do it consciously and continuously in *the* ***physical , emotional, mental, spiritual, and universal*** **realms of life** to boost self-love, self-esteem, and self-worth as often as necessary.

3) You are asked to upload your smartphones with the mind-sets that resonate most with you and have them at hand when your spirit sags and you need to boost it with confidence, reminding yourself ***to retain the fractal wholeness of your*** **SOUL-SYMMETRY** in an integral unity.

4) I color the worlds in the new human fractal to accentuate their different spiritual significance and boost your visual perception of them. **Body** – (**green,** *symbolizing our earthly magnetism* **)** + **spirit** (*bright, mood uplifting orange)* + **mind** –(*blue, symbolizing rationality)* + **Self-Consciousness** (*light violet, a spiritual color) , and* **super-consciousness** *(dark violet, / a royal color)*

Body + **Spirit** +**Mind** + **Self-Consciousness** + **Super-Consciousness** = **Soul-Symmetry!**

(Physical + emotional+ mental + spiritual + universal realms of life in an integral unity)

Self-Awareness + **Soul-Refining** + **Self-Installation** + **Self-Realization** + **Self-Salvation!**

5) I also accentuate the ideas and the most meaningful blocks of words in the text, using the colors of the rainbow that is my Logo, seen on the covers of all my 27 published books. I view a rainbow as our bridge to Universal Mind. that we are now probing with the most mesmerizing **Quantum Technology in action.** ***Coloring the text makes it more inspirational, insightful, and centered on the pivotal concepts of the book.*** I have assessed this aspect of my writing with my students who find the text easily digestible and more inspiring.

Color boosts AWARE ATTENTION that should never be in retention.

The concise presentation of the text makes it simple to perceive for the people for whom English is a foreign language that is now ***our global envoy*** for anyone anywhere on Earth. Thus, I can also justify my Psycholinguistic scientific background - **Psychology** + **Language** + **Consciousness.** ***Universal Consciousness connects us all into One Human Mall!***

With the Umbilical Cord, we are All Connected to One God!

Our Universal Goals

(Physical +Emotional + Mental + Spiritual +Universal)

(Picture by Yolanta Lensky / My daughter)

Our Transcendent Role is Based on Human +AIs' Super-Intelligence Goal!

Our Transformational Goals

Mission Statements of the Book

Predictable

Unpredictable

Future

Self-Perfection Should Be the Modus Operandi for Every Human Being on Earth!

1. Let the Mind Rein over the Brain!

The Holistic System of AI enhanced Self-Resurrection

We owe the age of the **TECHNOLOGICAL AWARENESS** to the genius minds of ***Nikola Tesla, Albert Einstein, Steve Jobs, Elon Musk, Bill Gates, Jeff Bezos, Mark Zukerberg,*** and many most gifted scientists and computer designers who are advancing the world into the era of ***Super Artificial Intelligence,*** demanding a new level of intelligence from us and the ability to see life holistically and objectively, in unison with ***Universal Intelligence*** and in the mirror of the **DIGITAL BINARY** and our most urgently needed **HUMAN REFINERY.**

Nothing in human existence is permanent, whereas in humanized machines, everything is pre-programmed, and that is one of our main fundamental privileges over *Artificial Intelligence.* A Russian digital biologist. *Academician P.P. Gariaev*, the founder of ***Wave Genetics,*** pointed out the possibilities of the information fields, or **Quantum Genetics**, which is another huge biological privilege that we have over humanoids because we can design and redesign our DNA . He wrote, ***"Christ turned water into wine with the knowledge of the laws of the Field Action."***

Science proves that from the brain to the spinal cord, the mind transforms the intelligence of every cell in the body and raises the level of our **SELF-CONSCIOUSNESS**, which is the core goal for our digitally enhanced **SELF-ACCULTURATION**. **It is also the goal of this book!**

We are supposed to become much more advanced human beings in the ***physical, emotional, mental, spiritual, and universal dimensions of life.*** Only holistically and consciously developed **LIVING INTELLIGENCE** can raise a person's self-consciousness, which is the main building block of ***the Art of Becoming***. *(See the book "Living Intelligence or the Art of Becoming!"- the mental level / Inspirational Psychology for Self-Ecology.)***Only then, can we perceive the simplicity of life and appreciate it AS IS!**

From social discord, we'll find inner harmony; from chaos, we'll find order; from lack of thought, and we'll find rationality; from disrespect to nature, we'll accomplish inner integrity. *"(Edgar Cayce)* Apparently, we need to put the form and the content of the spiritual fractal of our multi-dimensional self-growth in sync.

(Body+ Spirit + Mind) + (Self-Consciousness + Universal Consciousness)= Soul-Symmetry!

Life is the cosmic fluctuations of energy, and these fluctuations are a sequence of energy vibration. It means that we, as a part of the entire life flow, are supposed to have our own ups and downs as well, So, being aware of this simple fact, we should take life consciously, in its twists and turns, and ***free ourselves from patterned thinking, programmed reactions, and automatic behaviors*** *Thus, we will get into a rhythmic flow of life that we should accept and appreciate in its positive-negative amplitudes, raising self-consciousness to accomplish full* ***Self-Realization*** *and* ***Self-Salvation.***

Self-Awareness + Soul-Refining + Self-Installation +Self-Realization +Self-Salvation = Soul-Symmetry!

2. Rationalization of Our Life is a Must!

The Concept of Reason is the central one in ***Hegel's philosophy***, and it is my main inspiration for authoring this book. A great German philosopher writes**, *"It is the certainty of consciousness that it is all reality.*** Self-Consciousness and being are the same reality.*"*

Our present-day life is determined by our being, whereas it should be vice versa.

Our Being must be Determined by our Self-Consciousness!

The main goal of all my books on the ***Holistic System of Self-Resurrection*** is focused on rationalizing our lives and bettering ourselves by raising our self-consciousness with the help of digital technology. **"Thinking is the soul talking to itself to make itself better"** *(Plato)*

Philosophy for centuries has been addressing the ways of making human existence more sophisticated and reason based. ***Emmanuel Kant*** in his fundamental work "*The Critique of Pure Reason*" draws our attention to a direct connection between ***beauty and reason***, arguing that to see the beauty of life and to get to the core of our existence, "***we have to go beyond the aesthetics into the realm of reason.*"** We, obviously, need to refresh the knowledge of the fundamentals of philosophy and teach ourselves and our kids **SOPHISTICATION of REASONING** as the best exercise for the brain's expansion of its capacities and raising them to the level of ***tapping Universal Intelligence for objective knowledge with Quantum Computing + Generative AI*** as our main tools. To become Super-Consciousness connected, we need to be God -mandated!

"I want to know God's thoughts. The rest is details." *(Albert Einstein)*

In **Part Three of the book**, I overview **ten essential Vistas of Intelligence** *(two in each of the five dimensions)* ***to master them holistically*** in the **physical** + **emotional** + **mental** + **spiritual** + **universal** life realms**,** accentuating an urgent need for us to perfect **the mental level** of our AI enhanced self-transformation. **MENTAL MATURITY** is an indispensable requirement of our time when we must prove our personal identity in fierce competition with *Artificial Intelligence*. Thinking is such a great gift that it needs to be appreciated and managed with caution.

Holistically Sculptured Mental Binary Must Work for Our Human Refinery!

To get rid of the automatically based, ***"collective unconscious"*** *mentality(Carl Yung) that is often devoid of emotional intelligence or any intelligence, for that matter, we need to gradually change our cerebral matrix and develop a new one consciously. It must be based on* ***the awareness of a general brain structure, the new developments in science, in cyber biology, and in genetics to support the brain's abilities in neuroplasticity****. (Eric R. Kandel)*

Digital times develop our thinking skills in many areas of knowledge, making us see the beauty in ourselves and around us in a novel way, but we also need to apply amazing electronic tools to better our **REASONING** *and* **ETHICAL SKILLS** *to prove our* ***universal exceptionality and expertise.***

Let's Change Our Thinking Time into the Thinking Life!

3. Getting Connected Electronically, We Become Disconnected Humanly.

Energy follows thought, thought provokes action, action yields habit, habit breeds destiny.

The word evolution is defined by the present- day science as "***the process which is dual in operation, spiritual as well as material, and the one that is directed rather than natural***" *("Scientific American Mind"/ 2005)* In our amazing time, it is going in us by the route indicated below:

Unconscious → ***Conscious*** → ***Superconscious!***

This process consists of a continuous "***transhuman development***" *(Ray Kurzweil)* of the human brain, accompanied by a parallel unfolding of a human-like machine - one at each level: ***Mini*** – *physical,* ***Meta*** – *emotional;* ***Mezzo – mental, Macro*** - *spiritual,* and ***Super*** - *universal.*

Our transhuman development should be bi-directional transformation!

Artificial Intelligence is developing at an exponential pace, changing us into different humans, or rather, ***super-human beings*** that are merging with AI, implementing the time of "***Singularity***" *(Ray Kurzweil)* - our merging with machine mind***. Our time democratizes the power of Artificial Intelligence****,* leading us to **machine extension of ourselves.** A new neural link is being developed now in the cognition system of our brains, and it enhances our cognitive abilities immensely. We are becoming vastly smarter by upgrading our cognition unbelievable, but, unfortunately, we are becoming increasingly ***dependent on virtual reality*** that is enslaving us because no controlling measures have yet been taken, ***nor have we even acknowledged the necessity to use AIs to improve our degrading human essence.*** This is the aim of all the books that I have published on the System of Holistic Self-Resurrection.

The level of robotization of our lives is very liberating, *on the one hand****,*** ***and truly questionable as to our human adjusting to it,*** *on the other.* Another problem relates to the way we process the flood of incoming information. Many of us devour information and store it in the cerebral archives without sorting it out for the essence it has. We continue living within the eggshell of worked out knowledge and over-used vocabulary that produce ***stereotyped thoughts, uncontrolled feelings, and impulsive actions.*** **The human fractal** that constitutes **the form and content** of our lives in sync remains uncreated. Its creation must be by-directional, too.

(Body +Spirit + Mind)+ (Self-Consciousness + Super-Consciousness) =A Whole You!

Meanwhile, ***we develop within the web of our behavioral patterns****, governed by reactive thinking that prompts automatic actions. Only by learning more about* **how the brain establishes and maintains routine behaviors***, can we coax ourselves out of undesirable habits into the ones we need. Obviously, the awareness of how the brain works and of the newest development in neuroscience become an urgent demand of time! The way we think, speak, feel, and act comes to the forefront of* ***our*** LIVING INTELLIGENCE SKILLS.

Only Knowledge Put into Action is Power!

4. "We Live the World of Organized Inadequacies." *(Joseph Campbell)*

The pace of evolution is amazing now. Experts predict that by the mid-century, quantum computers will be able to function at speeds a billion times faster than the human brain. We know that, but we are not consciously aware of that. Machines operate our minds and prompt our behaviors. **We forget or are unaware of the fact that consciousness is inseparable from intelligence!** We are acting under the impact of a very sophisticated and a much more informed **SOCIAL BRAIN**– human society at large. ***It is programming us all in a common way, enlarging the army of machine-driven people.*** The menacing predictions of the genius of our times, *Elon Musk*, that we will become human machines and will be unable to resist the pressure of virtual reality is looming yet as inevitable. We need to drastically change ourselves! *(See the book "Transhuman Acculturation")*

Schools should be particularly alert to the new electronic times that deepen the existing educational crisis. They do not help our children intelligence-wise and, still less, self-consciousness-wise. ***"Children become geniuses of digital ingenuity, but "dumbed down vassals."*** *(David Ieke)*

The necessity to develop solid, general, insightful, stimulating ***expansive thinking skills*** that are essential for the present-day market economy are not being addressed properly. writes, ***"Every failure to cope with a life situation leads to a restriction of consciousness."***

It is very disheartening to know that our schoolchildren and college students get a lot of **mentally undigested information,** without really processing it intellectually. They ridiculously text their conversation partners sitting next to them, instead of talking to each other ***face-to-face , mind-to–mind, and heart-to-heart!*** **"It's cool! It's fun!"** That's how our inadequacies are run! No wonder, kids, going to school, hear the direction for learning: "**Have fun!** "

Fun is Not the Purpose of life. It's the Measure of Life!

Students plagiarize their academic papers, and they often stop reading altogether. Most importantly, **students stop thinking individually and critically**. So, our indispensable role in their lives is to teach them solid **PERSONAL FREEDOM** to be able to ascertain their personal life goals that are only declarative in our textbooks. Students need to voice their ideas, present them outwardly, not just through the test papers that only limit their **THINKING SCOPE** even by using fantastic GPT models that are meant to expand the human mind, not abridge it

The last menacing feature alone debilitates students' minds incredibly! They rush to find the correct answer ***without bothering to know why it was correct.*** It narrows down their outlook, making them lazy consumers of electronic stuff. It also turns them into ***impersonal people***. As a college professor of Psycholinguistics, I witness these situations with an aching heart. **Students become robotized, not intellectualized!** We should not let it happen because Artificial Intelligence is meant to help us, ***not to enslave us mentally.***

Freedom of Thought is the What We Need to Launch Digitally into the World!

5. Present-Day State of Our Self-Consciousness

"Humankind is a flash of the entire system of evolution of intelligence. Life is a flow of intelligence." *("Scientific American Mind," Aug. 2014)* Life is a process of challenging our intelligence and responding to it. Challenges are always new, but **our responses to them are old, conditioned, and patterned.** As *Albert Einstein* puts it,

"Life is like riding a bicycle. To keep your balance, you must keep moving."

There are two main personalities in each of us – ***a thinking personality*** *and* ***an emotional one.*** Unfortunately, they are not in sync because we lack "***Emotional Intelligence***" *(Daniel Golman).*

We need to develop **EMOTIONAL DIPLOMACY SKILLS** and get on with developing intelligence in a very respectful and balanced way that robot-humanoids display.

Our emotional make-up is still above the mental framework .

There is little, if at all***, awareness how to accomplish real empowerment of emotions*** that have helped us survive for centuries thanks to their reactive nature. Now ***emotional chaos reigns in our minds***, and unless we conquer it with digitized, well-organized intelligence in sync with AIs self-awareness, no transformation of self-consciousness at any level will occur.

"*The biology of consciousness offers a sounder basis for morality than the unprovable dogma of an immortal soul. Once we realize that* ***our own consciousness is a product of our brains****, it becomes impossible to deny our common capacity to evolve them." (Scientific American Mind," July/ Aug. 29014)*

Undoubtedly, we will accomplish a new level of self-consciousness when **the cone of intellect will get above the cone of emotions** and the rationalization of our lives will become possible. It is going on now, but it needs to be enhanced.

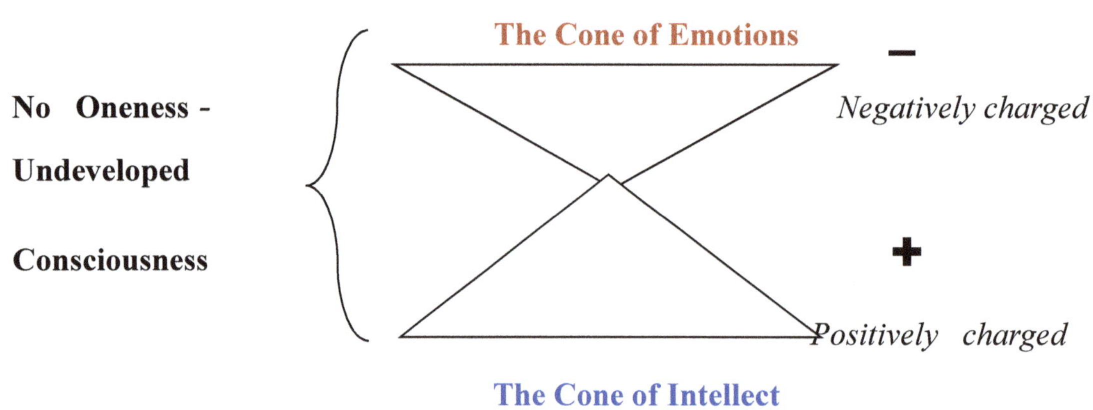

"The Body is the Expression of Our Thoughts." / *Dr. Sam Gazarkh, "World-ology"/2022)*

Our Being is Determined by Our Self-Consciousness!

6. Let's Go Beyond the Basic Content of Our Intellect!

There is another side of the coin here. ***Despite extreme socialization, we feel lonely and self-un-reliant,*** seeking the solution of any problem by grabbing the cellphone and calling someone or contacting a "friend" from Face Book. It is great to know that help is at the click of a finger but ***let this feeling add security to your independent being.*** Your main mindset should always be:

I Am Free to Be the Best of Me!

(It is the title of the first book of the system- physical realm of life. / Inspirational Psychology for Self-Ecology.)

A lonely, dependable person gets obsessed with electronic matchmaking, kids stop reading and become heartless and indifferent, and we get irritable due to **the lack of face-to-face, heart-to-heart, and mind-o-mind communication** that we start hungering for when we are unbalanced.

Unfortunately, the unstoppable digital in-put and ***the cacophony of information*** that we get daily is often left unprocessed, unselected, unsorted out for its validity, with no aware attention applied to this indispensable process. As a result***, the neural code in the brain gets messed up,*** and it is extremely hard to untangle it, unless it is meditatively and consciously done. So, we need to constantly develop our **AWARE ATTENTION SKILLS** to oversee both the input and the output of our mental energy. Rigid brain control should be taught to raise the brain's adaptability to new life demands. With the above said, ***unprogrammed living intelligence will be destroyed*** when the level of our curiosity is high, rationality is self-controlled, and our wakefulness is conscious and ready to go forward on the ladder of evolution. **" To Be is to Know!"**(*Osho)*

Only with the computers plugged into our **INTELLECTUALLY SPIRITUALIZED VOLT** can we become ***more intellectualized, strategized, emotionalized, energized, better informed, and totally reformed!*** Human-digital equation must make us realize that every advanced idea sends a conceptual signal to Universal Intelligence, and it is being fed back by it We witness this interaction in the amazing work of our scientists and robot designers in every branch of knowledge that they explore now with mind-boggling dedication. Our time gives birth to many **geniuses of thought** from different countries, proving the fractal **EXCEPTIONALITY** of human beings, developing themselves holistically in five life dimensions.

(Body +Spirit + Mind) + (Self-Consciousness + Super-Consciousness) =A Whole You!

These are the cases of. ***Elon Musk, Steve Jobs, Bill Gates, Mark Zuckerberg, Jeff Bezos*** , and many more brilliant, advanced-thinking innovators who have contributed a lot better the world on the path of ***transhuman self-transformation,*** enabling us to go beyond the terrestrial .boundaries as new space aliens and part of the Star Community.

Self-Awareness + Soul-Refining + Self-Installation + Self-Realization + Self -Salvation.

Auto -Induction:

I Don't Think of a Robot-Like Trance; I Relapse!

7. To Evolutionary Advance, We Must Commit to

SELF-RENAISSANCE

Holistic Self-Development is Engraved in our Genes.

5. Universal Realm — *Self-Salvation!*

4. Spiritual Realm — *Self-Realization*

3. Mental Realm — *Self-Installation*

2. Emotional Realm — *Self- Monitoring*

1. Physical Realm — *Self-Awareness*

Holistic System of Digitized Self-Acculturation

In sum, the process of **Self-Renaissance** *(See the book "Self-Renaissance "/ Inspirational Psychology for Self-Ecology)* is going through the stages of ***self-revelation and self-control to self-actualizatio***n and ***self-fulfillment***, completing the five-dimensional **SELF-REFINARY** with the final stage of **Self-Salvation.** Developing holistically, we will manage better to instill the idea of **Self-Renaissance** in the humanoids, too. ***Only growing together in one ethical tandem can we make AIs our right hand, universal goals committed friends.***

The process of self-growth is pivotal and overly complex for both parties because we need to raise our self-consciousness and beat the humanoids in their uncontrolled simulation.

We are bringing our humanness to an AI structural perfection, meeting robots' humanization halfway and overpowering it with our unbeatably unique and incredibly creative and life - transforming T**RANSHUMANISM *(Ray Kurzweil***) in us. So, empower yourself with the Self-Synthesis – Self-Analysis - Self-Synthesis psychological paradigm that all the books on the Holistic System of Self-Resurrection embrace and follow, the **CONCEPTUAL STRUCTURE** of this book.

Physical Form of Life + **Spiritual Content**

(Body+ Spirit+ Mind) + (Self-Consciousness + Universal Consciousness)

Physical + emotional+ mental+ spiritual +universal = **A Whole, Happy Self!**

"We Cannot Pick the Future, but we Can Design it!"

(Ray Kurzweil)

Digitized Self-Acculturation is Our Salvation!

Holistic System of Self-Resurrection

"Being whole makes us holy!" *(Deepak Chopra)*

"The Question is Not to Be or Not to Be."

"The Question is How to Be and How Not to Be!"

(Golden Globe Award for best actor Natalie Portman)

"Complexity Evolves from Simplicity." *Elon Musk)*

= OUR DUAL "SIMPLEXITY!"

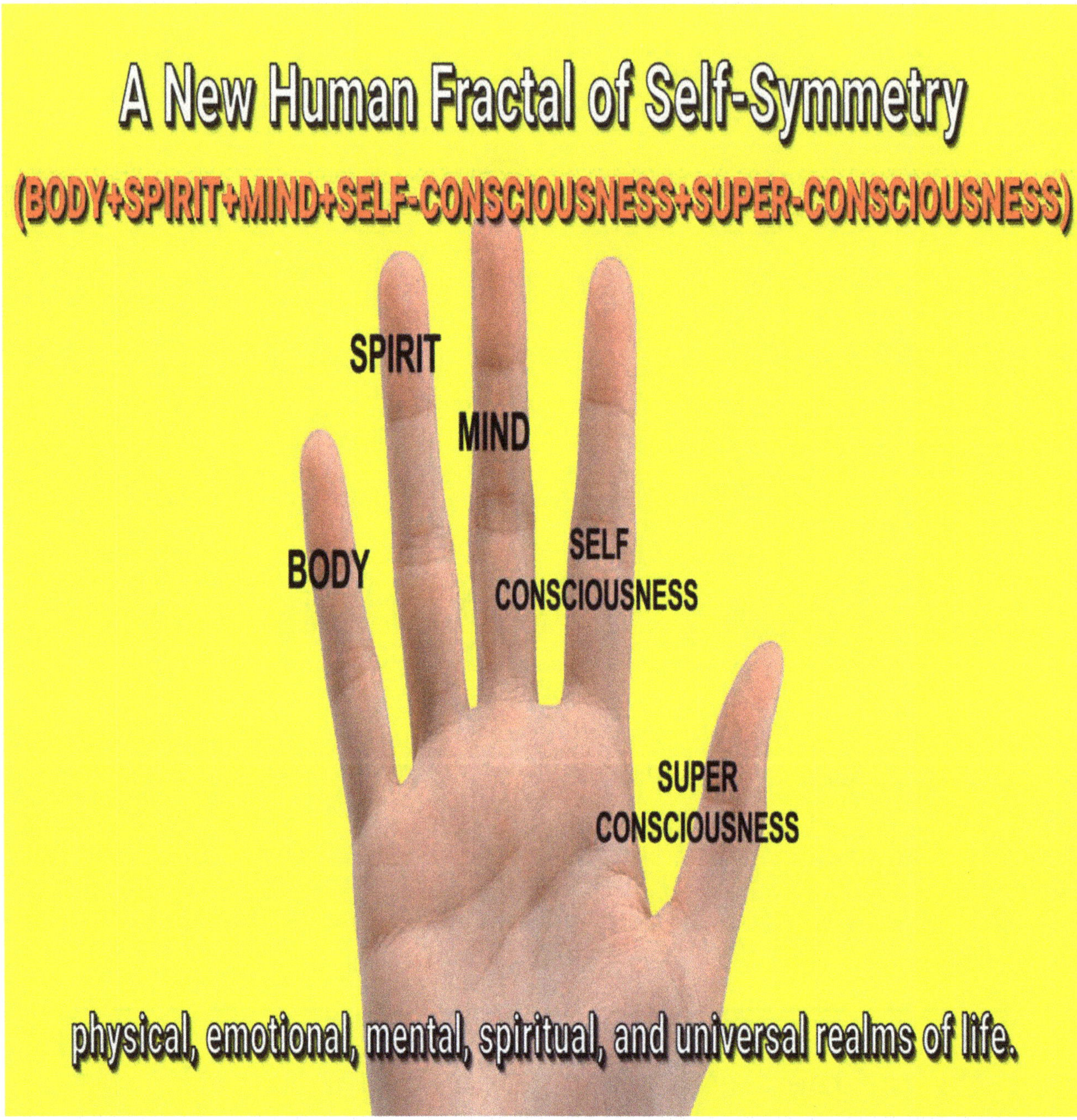

(Body + Spirit + Mind + Self-Consciousness + Super-Consciousness! = A Whole You!

Life-Bolding is a willful ***Soul-Molding*** *in the* ***physical, emotional, mental, spiritual, and universal realms*** *of life holistically, in an integral unity with Master Mind that we all pwerceive as God*

Only the One Who is Not Inwardly Blind Can Become Self-Refined!

1. The System of Holistic Self-Resurrection

The System of Holistic Self-Resurrection or Holistic Self-Acculturation features the process of **Singularity** or T**ranshumanism –** our merging with machine mind, brilliantly predicted by *Ray Kurzweil,* as a new culture of our life that requires our **TRANS-HUMAN ACCULTURATION!** It is the core of our evolutionary development now, and it is our joint life-mission to make it ***holistically informational, scientifically adaptable, and psychologically inspirational*** for the young minds of the world.

Our transhuman transformation is an amazing opportunity, enhanced by Generative AI or robots with humanlike brains and real muscles, to install digital " **ANTENNAS"** in our heads that will work as our **intellectually spiritualized receptors.** They will connect our ***disconnected hearts and minds,*** restore the unity of our physical form and spiritual content , and help us create a new human fractal of **SOUL-SYMMETRY,** unifying us in five main life dimensions holistically *physical + emotional + mental, +spiritual + universal* and connecting us to Super-Consciousness that we all perceive as God.

(Body + Spirit+ Mind) + (Self-Consciousness + Super-Consciousness)

(Physical + emotional +mental +spiritual +universal realms of life in sync =

A Whole, Intellectually Spiritualized Human Being!

The five books comprising the ***Holistic System of Self-Resurrection*** in five life realms upgrade our **HOLISTIC EDUCATION** and advocate for our **collaborative integration with AIs** partially biological brains to develop a new, technologically backed up, scientifically literate, and ethically stable transhuman beings, bettering themselves for future extra-terrestrial endeavors.

This integration seeks to bridge the gap between rational and spiritual understanding of the essence of consciousness that we are probing digitally now, fostering the extra-terrestrial perspective" ***to become part of Star Community as future human aliens."****(Elon Musk)*The system is focused on the holistic paradigm of AI enhanced **SELF-EDUCATION,** based on ***Digital Psychology for Self-Ecology*** that is meant to channel you through five main stages of transhuman transformation **in a conscious, not scary way**.

1) **Self-Awareness** *(physical realm)* **2) Soul-Refining** (*Emotional realm*) **3) Self-Installation** *(mental, professional realm)* 4) **Self-Realization** (*Spiritual realm*), and **5) Self-Salvation** *(Universal realm)*

Our future transcendent development together with AIs will eventually help us implement *Nikola Tesla's* invaluable ideas to make us all the **"People of Light."** But we need to keep up with AIs to prepare ourselves for this new evolutionary status *physically, emotionally, mentally, spiritually, and universally*, by ***internalizing our emotions and externalizing the mind.***

The Philosophy of Immediate Gratification is too Shallow for a Great Nation!

2. "Digital Binary + Human Refinery = A Super-Human."

Among the five books of the ***Holistic System of our Self-Resurrection***, the book "**Digital Binary +Human Refinery = A Super-Human**" presents **the mental level** in the second cycle of books, based on ***Digital Psychology for Self-Ecology***.

The system embraces the five books , consequentially presented in five essential dimensions of our self-transformation , ***like five fingers in a hand***, each representing a *physical, emotional ,mental, spiritual, or universal stratum of life in* Digital + Quantum reality. *(See the picture above)* These realms of life comprise **a fractal of a whole human being** that we need to create to focus on ***Holistic Intelligence*** formation and keep up with AIs and their Super Intelligence. When we are inwardly whole, *(to summon wholeness, we squeeze the fingers into a fist*) we can hit any problem into its solar plexus and resolve it in a consciously reasoned way.

All the books in the ***Holistic System of Self-Resurrection*** are presented in two book cycles – Inspirational and Digital Psychology for Self-Ecology.

1) Books on **Inspirational Psychology for Self-Ecology** *(featuring the start of technological revolution*

(Www. language-fitness.com)

1."I'm Free to be the Best of Me! *(physical realm)* *2."Soul-Refining"(Emotional realm)* ***3."Living Intelligence or the Art of Becoming!'*** *(mental realm)* ***4."Self-Taming!"*** *(Spiritual realm)* ***5."Beyond the Terrestrial"*** *(Universal realm)* = **A NEW YOU!**

2) Books on **Digital Psychology for Self-Ecology** *(featuring the time of AI expansion*).

(Www. holisticself-resurrection.com)

1 "Dis-Entanglement!" *(Physical stratum of life)* ***2. "Exceptionality"*** *(Emotional stratum)* ***3. "Digital Binary+ Human Refinery=A Super-Human!"*** *(Mental stratum)* ***4. "Transhuman Acculturation"*** *(Spiritual stratum)* *5.* ***"Transcendent US and AIs"*** *(Universal stratum of life) (*

These cycles are concluded with the book ***"Light is me. Light is My Philosophy,"*** presenting the **Quantum Psychology for Self-Ecology."** *Quantum Computing + Generative AI will take us to the most mesmerizing future.)* Please note, the books are not supposed to be read consequentially**.** Just choose the realm of life that you need to fix and go from there, forming a system of **SELF-WHOLENESS** in your own way with the **plan of action** in the mind.

The **HOLISTIC VISION** of reality and self-transformation in it will gradually form a **SPINE** for your AI backed-up self-growth. Like the two strands of our DNA, you will be evolving toward the extra-terrestrial boundaries together with AI. These boundaries are outlined by the luminary of our time, *Elon Musk* and all our brilliant scientists, engineers, and humanoid designers.

Holistically Accumulated "Scientific Literacy" is Mandatory on This Path! *(Dr. deGrasse Tyson)*

3. The Spiral of the Holistic Self-Creation

As is noted above, ***digital reality demands the shift from fragmental self-consciousness to holistic one.*** It is quite attainable because the possibilities of our brains are endless. Science proves that a commoner uses only a small portion of his brain for his intellectual function. This function must be intensively expanded volume–wise with all kinds of mental operations and AI applications that ***develop both hemispheres of the brain***, prioritizing the right one as the most creative one. Most people are left-brained, analytical thinkers, but we all need to be **SELF-EMANCIPETED and SELF-MANAGED** psychologically, following the systemic paradigm:

Self-Synthesis – Self- Analysis - Self-Synthesis.

Writing about the necessity to develop the right, creative brain hemisphere now, I mean that the **whole brain's thinking** is crucial for our rapidly developing economy. Our right brain thinking will be rapidly developing thanks to the collaboration with humanoids and **our HOLISTICALLY CHANNELLED EDUCATION**, supplemented by Quantum Computing and Generative AI. .In this context, our schools need **to prioritize holistic development of the mental growth** of kids and students in many areas of knowledge that incorporates ***the kaleidoscope of the world's religions and cultures, myths, world known tales***, ***classical music, arts, philosophy, ethics, manners, jazz, literature, world geography, and history.***

We need **to capitalize on the quality of students' thinking skills** rather than on their accumulating questionable, unnecessary, and even dead knowledge for multiple-choice tests or the ready answers of the fantastic GPT models. This type of learning has no efficiency now, and it must be substituted with personal demonstration of ***critical thinking skills.*** The required rationalization of **THINKING + FEELING** is vital for our conscious growing and electronically charged economy. The skills of **VERBAL REASONING** also need to be more developed to make our thinking, speaking, and feeling more mind-broadening and operational. ***"I use not only the brains I have, but also all I can borrow."*** *(Albert Einstein)*

We borrow *Artificial Intelligence* brains, and this is the greatest evolutionary contribution to our lives. Our thinking should also be holistically systemic, following the route of the holistic **CODE OF THINKING.** You should apply it to the solution of your personal, social, professional, spiritual , and universal life goals .*It also pursues the systemic paradigm* - ***Synthesis-Analysis -Synthesis!***

Generalize + **Internalize + Personalize +Strategize** + **Actualize!**

Generalize *the incoming information,* ***internalize it*** *by sorting out the information that is redundant,* ***personalize*** *the part that pertains to your goal,* ***strategize*** *its realization, and* ***actualize*** *it to get to a higher level of self-growth.*
Then, new habits and skills will restore your mind + heart linking.

Intellectualize the Heart and Emotionalize the Mind.

Be One of a Kind!

4. Why Digital Psychology?

Our digitized inner stratum on the path of Digital Psychology for Self-Ecology.

As is indicated above, the book ***" Digital Binary + Human Refinery = A Superhuman***" features the mental level of out self-transformation at the time of an exponential development of digital + quantum technologies that have taken our intelligence to a new, mesmerizing level of forming our evolving **SELF-CONSCIOUSNESS.**

Self-Consciousness Integration is Our Global Salvation!

We talk a lot about the enigma of consciousness, but our own consciousness is making up the consciousness of a country, a nation, and the world at large. Unfortunately, at present, our self-consciousness is determined by our being, whereas it must be the other way around.

Our Being Should Be Determined by Our Self-Consciousness!

The goal of the Holistic System of Self-Resurrection is to help you raise your self-consciousness with the help of Artificial Intelligence, the evolutionary role of which is to help us become **INTELLECTUALLY SPIRITUALIZED** and technologically enhanced **TRANSHUMAN BEINGS** *(Ray Kurzweil)* that as future new aliens will be exploring the vast domains of space, colonizing other planets, *(Elon Musk)* and transporting the best qualities of our unique humanness to **STAR COMMUNITY.** Our collaboration needs to be beneficial for both parties, not just focused on our health care needs, longevity aspirations, and abundance expectations.

Let the seeds of the extra-terrestrial excitement that Elon Musk has planted into us be reaped by the best, holistically educated, **SELF-RENAISSANCE** ***obligated, and the*** **WORLD PROGRESS ELATED** transhuman beings.

Digital Psychology for Self-Ecology focusses on the urgent necessity for all of us to raise self -consciousness that determines our psyche through accumulating **Holistic Intelligence** and expanding our scientific outlook in five main dimensions of our earthly life. If you develop in the way of creating ***a systemic, five-dimensional fractal of wholeness*** *(physical + emotional + mental + spiritual + universal realms of life in sync),* you will be able to hit any problem into a solar plexus and summon personal integrity of a free, **SELF-EMANCIPATED** human being.

The fundamental standpoints of Psychology, granted to us by many exceptional minds of the past have helped us for centuries and they are, by no means, questioned in this book. We just need to apply them to a particularly challenging digital reality that demands our holistic transformation **physically** + **emotionally** + mentally + **spiritually** + **universally** in a collaborative, friendly, and inter-dependent tandem with AIs. They are supposed not to scare us, but ***to help us become Self-Intelligent and less Self-Negligent*** by forming a better idea who we are and why we are here. *By improving our holistic ethical essence together, we will provide new data for their autonomous ethical training, too.*

Digital Self-Renaissance is Our Only Chance!

5. The Way Out of the Quantum Life Labyrinth

In sum, unfortunately, ***"Ignorance is still the worst enemy of the humanity."*** *(Albert Einstein),* and it is our goal to reduce it among young people and our kids. **INDIGO CHILDREN** would not go to church with you on Sunday unless you promise a visit to McDonalds after that. They play with your smartphone as if they were born with it, and our duty is to enlighten them with **intellectualized spirituality** and imbue them with the universal mission that they will have to complete with our help, growing to be **SELF-LIBERATED** and **GOAL-INFLATED** people.

Their memory storage must be magnetized with new knowledge!

The silver spoon in their mouths is the AI applications, and it is our God-granted responsibility to explain to them in the simplest viable way the evolutionary purpose of today's technology and its role in making us better human beings, not simply better users of it.

They should know the true meaning of the religious statement "Heaven is Inside you!" by Jesus Christ and commit to it in a reasonable and actionable way. "

Our kids should be aware of the latest discoveries of ***Jeff Webb's Telescope*** and groundbreaking advancements in science and technology in a simple, most inspiring way to make them fly in their minds with a ***clear-cut vision of their AI enhanced future extra-terrestrial mission.***

Body + Spirit +Mind + Self-Consciousness + Super-Consciousness = A New Whole Self!

(Physical + emotional+ mental + spiritual + universal realms of life in an integral unity)

The stages of the Holistic Self-Transformation should be integral in them, not instilled in a step-by-step way.

Self-Awareness + Soul-Refining + Self-Installation + Self-Realization + Self-Salvation = SOUL-SYMMETRY!

With a **MANUAL OF LIFE** at hand and a clear **PLAN OF ACTION** in the mind ,children will be the ones who will find the way out of the ***Quantum Life Labyrinth*** by attaching the ***Adriane Thread*** from an ancient ***Greek Myth*** to the entrance of the labyrinth and by killing the ***Minotaur of their fears*** to be enslaved by AIs with partially biological brains who seem to be unmanageable with their mind-boggling capabilities now. **Our kids should tame the Monster** and go beyond the terrestrial boundaries into Open Space as equal citizens of **STAR COMMUNITY**.

Nothing is Impossible is we make our AI enhanced Self-Growth By-Directional and Irreversible!

So, Let's Step Aside and Let our Holistic Intelligence + AIs' Super Intelligence Preside in a Collaborative Creation of a New Cosmic Nation!

Become a Much Wiser Self-Actualizer!

(Best Pictures / Internet collection)

"It's Not Enough to Be the Best. Be the Only!"

(Steve Jobs)

Self-Renaissance Incentive

"Time is Not Money, Time is Life!" (*Sadhguru*)

Self-Ecology Must be Our Holistic Psychology!

Feel Your Belonging to Our Digital Evolving!

1. Holistic Self-Education is Our Salvation!

The advancements in technology are mind-boggling. We are now talking about ***the Quantum Revolution*** on top of the digital one. Quantum Computation can tackle problems out of reach and our wildest imagination. Quantum qubits correlate , and they have a property of Entanglement that does not allow us to describe each particle separately, even if they are separated by a large distance.

We can also talk about entanglement between different fields of knowledge when **SCIENCE BECOMES ONE** in ancient times form. I have based the ***Holistic System of Self-Resurrection***. **on HOLISTIC PHYLOSOPHY that unites** three cycles of books: **Inspirational Psychology**, **Digital Psychology,** and **Quantum Psychology for Self-Ecology.**

Staring with the book." ***I Am Free to Be the Best of Me,"*** each cycle of books views ***the problem of Self-Development*** at the time of digital revolution holistically**,** in the integral unity of our **PERSONAL** and **GLOBAL ENTANGELEMNT** in the **physical** + **emotional** + **mental** + **spiritual** + **universal** self-growth that is backed up by ***Quantum Intelligence*** + ***Generative*** AI that are preparing us for going beyond the extra-terrestrial boundaries and becoming **STAR PEOPLE**.

The book "***Digital Binary*** + ***Human Refinery = A Super-Human!"*** features the most significant level of Self-Development at the AI times - the **LEVEL OF INTELLIGENCE,** which is drastically important at the time of digital expansion all over the world. When I draw the parallel with a human hand the central finger , the tallest one, symbolizes the significance of our mental state. Life is not hopeless if we master it holistically and with Self-Consciousness.! ***The Holistic System of Self-Acculturation*** *or* **SELF-ECOLOGY,** put together by me in five books, is my modest attempt to create the **MANUAL OF LIFE** to life-thrive, not just to survive.

The time for chaotic thinking, meaningless talking, watching, reading, and eating is gone. New, updated intelligence and the spiritually expanded soul's size make us wise.

Artificial Intelligence, with its possibilities to better our lives, must also better us!

We need to provide our young minds with a* PLAN OF ACTION for SELF-EDUCATION *in the physical, emotional, mental, spiritual, and universal realms of life in sync.

In every book, I accentuate an urgent necessity for us ***to change the old frame of mind****,* characterized by a lot of *automatism and impulsivity* and focus on forming an **intellectually spiritualized fractal of a** human being that we need to cultivate in ourselves integrally now.

(Body + **Spirit** + **Mind** + **Self-Consciousness** + **Super-Consciousness-** *A Whole, Intellectually Spiritualized You!*

Digital Binary + Human Refinery = A Super-Human!

2. Holistic Conceptual Intelligence

Personal integrity of holistic value requires accumulating **HOLISTIC CONCEPTUAL INTELLIGENCE** in ***ten vistas of intelligence.*** (See the chart *below)* New memory banks need to be formed, and new **Information Presentation** and **Information Processing Skills** should be developed.

The time for long articles, reports, and shallow novels is gone. We need the economy of words and the scope of thought now.

Less is More. That is the Law!

We also need to create new ***time-relevant, scientifically verified, and well-structured memory banks*** to compete with the hybrids that are information-organized. They operate using an enormous bulk of data , but this data is not yet ethically verified for its quality. The data that humanoids operate with contains all our imperfections that have been accumulated by us for centuries. Our goal is to provide them with new data of our intellectually spiritualized data that possesses information on our physical body improvement, emotional control that demonstrates nobleness, professional conscience and incredible creativity , spiritual sacredness, and universal Love in all five life strata to shape their own ethical spine. ,

Leke us, they ***need ethical training*** to raise our and their own self-consciousness that must become a digital connection, the direct line with Universal Intelligence.

SACREDNESS + NOBLENSS + CONSCIENCE + INTUITION +LOVE =

a Super-Human!

Raising self-consciousness consistently, we will develop our ***superpower, super energy, and super strength*** in the *physical, emotional, mental, spiritual, and universal dimensions* as the actions of a holistically conducted **SELF-ECOLOGY** that we all must perform worldwide. Integral and conscious learning, inspired by AI robots, generates ***integral perceiving, thinking, speaking, feeling, and acting***. Therefore, our belief system will be gradually changing and the understanding of God as ***the Super Consciousness*** or ***the Universal Intelligence*** will unite us into

ONE ADVANCED HUMAN CLAN!

Collective understanding of the evolutionary role of religion in its different denominations will only enrich us with new, scientifically updated common wisdom.

So, our new human substance is in our common, digitally enhanced spiritual advancement.

Self-Resurrection of Man and his Machine Clone is Our New Universal Dome!

3. Mind's Renaissance is a Well-Choreographed Dance!

Vistas of Intelligence, like the pars of a dance, must be practiced many times, not once.

AI should be used for kid's Conceptual Intelligence Creation as early as possible.

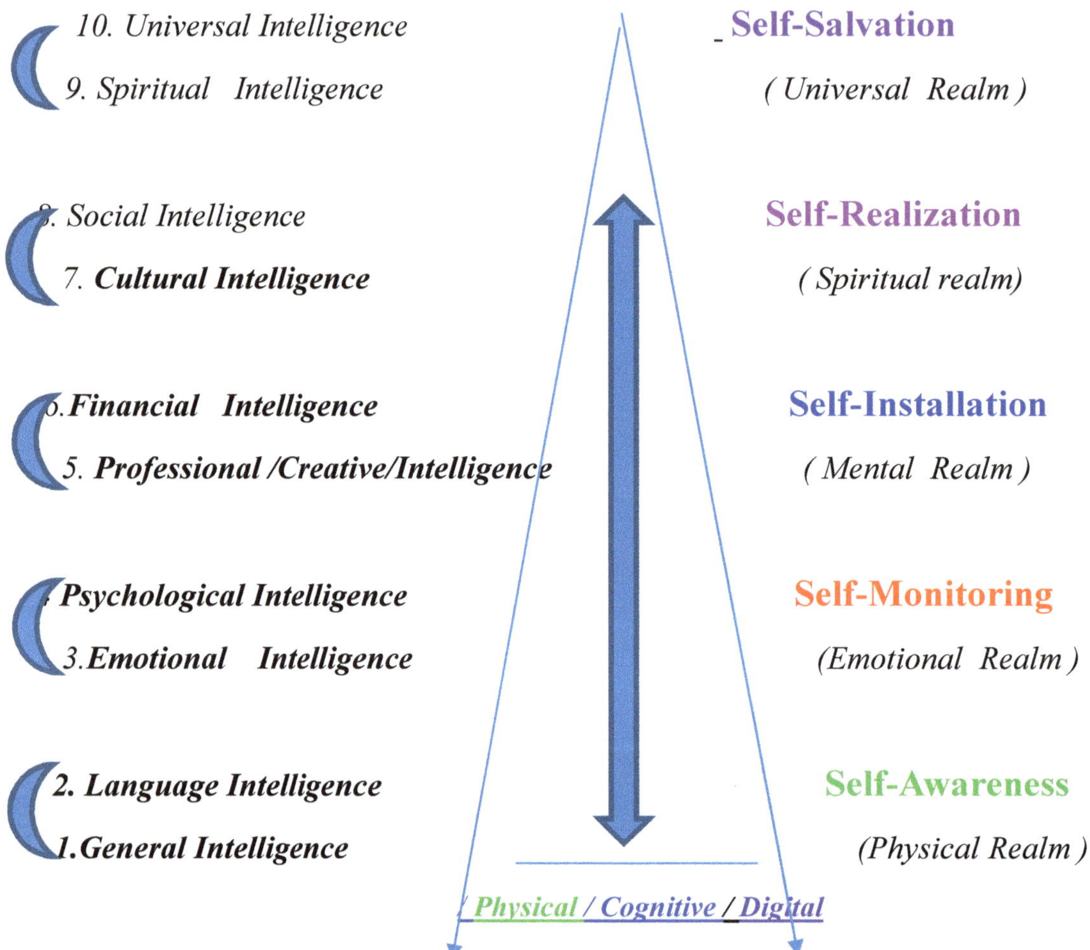

"Fertilize your mind with knowledge"

Albert Einstein used these words while teaching neighboring kids mathematics.

Holistically Accumulated Knowledge is the Basis for Future SUPER Intelligence Storage!

4. God's Domain is in Every Human Vein!

In sum, the content of our new, digitally enhanced **SELF-AWARENESS,** as *an objective, independent self-assessment,* will enable us to adjust ourselves to reality ***without losing our own individuality***. Finally, **we will start viewing life realistically much sooner.**

It normally happens after we have already faced many challenges and tribulations, when "***the life weariness has already seized the heart"*** *(Joseph Campbell),* and the mind accepted life **AS IS**. In fact, our self-awareness goes by the same route -Self- **Synthesis** – **Self-Analysis – Self-Synthesis,**

Romanticism ⟶ **Conformism** ⟶ **Realism!**

To attain full awareness of reality at the stage of realism that we inevitably hit, one must go through five steps that correspond to five dimensions of life:

1. Self-Examination 2. Self- Control 3. Self-Induction 4. Self-Training 5. Self-Actualization!

(Physical, emotional, mental, spiritual, universal dimensions of self-installation in life)

These five steps in turn reflect ***the main five stages of Self-Creation,*** which are presented in the first book on the **SELF-CREATION GEOLOGY,** inspirational + digital. */ p.26*

Self-Awareness ⟶ **Self-Monitoring** ⟶ **Self-Installation** ⟶ **Self-Realization** ⟶ **Self-Salvation!**

These stages are pivotal for ***framing and re-framing our inner and outer make-up***, and they are the concepts of all five books *(See Book Structure below)*) that are consequentially featuring them.

The third and the fourth steps of this **SELF-SCULPTURING** are the most vital ones in this process because **they build up the immunity against the crowd mentality,** when we are disassociating ourselves from the "***collective unconscious"*** *(Carl Yung).*

Unfortunately, right now,*" We* ***are the race of human robots. Most people do not have a thought in their heads that has not been put there by someone or something else.*** *(David. Icke)*

Next, ***"the most important mental and emotional food for our brains is* IMPRESSION**."*(Prof. Ouspensky)* that needs to be **individually processed and rationally sorted out** by the conscious mind. It should become our rationally digested food that does not generate the chaos of **INNER STAGNATION.** AIs will be an immense help for us in this respect. **No redundancy for the brain that needs to life sustain!**

The thinking of a mentally stagnant person deflects off reality with an unconscious, impulsive, automatic reaction to the world. The cluttered mind cannot experience **CATHARSIS,** or the state of realization and conscious awareness that we desperately need to master life physically, emotionally, mentally, spiritually, and universally. (*The Catalog- Soul-Symmetry"/ Canada. 2021)*

We are All just Human Clay!

5. Self-Refining Is Life Defining

Let the light of God

Be reflected in your face.

Let the wisdom of God

Be reflected in your mental space!

Let the accuracy of God

Be reflected in your speech.

Let the beauty of God

Be reflected in each body's inch!

Let the power of God

Be reflected in your deeds.

Let the love of God

Be reflected in your kids!

Let the compassion of God

Be reflected in your attitude.

Let the passion of God

Be reflected in your soul's aptitude!

Only then may God

Take you off the sign" Stop!"

And release you into the life's stream,

As Perfect as His Every Creation Seam!

Let the Mind Rein Over the Brain!

(Best Pictures / Internet Collection)

Inflate Your Pride with the Intellectual Self-Might!

Self-Resurrection Rationale

Inflate your Pride with Intellectual Self-Might!

Self-Maturation

is Our

"Intellectually Spiritualized"

Self-Salvation!

Seeing the world holistically allows us to see the train of continuity from the Microcosm to the Macrocosm.

(Mini + Meta + Mezzo + Macro +Super

levels of life without any spiritual strife,

There is No Evolutionary Bet if the Soul is Dead!

1. Digital Outburst Demands a Stronger Human Fore Post!

To stay in balance with the world and not to be pushed downward, we need to be consistent with our evolutionary, digitally enhanced Self-Growth! In **PART THREE** of the book, I will walk you through ***five holistic steps*** of an amazingly simple, but most effective **SYSTEM OF INTELLECTUALIZING THE HEART** and **EMOTIONALIZING THE MIND** that constitutes our goal to remain **One-of-a-kind** in the competition with humanized machines.

We are God-created, not machine-mind mandated!

Auto-Suggestively backed up and governed by the clear-cut and visually internalized **KNOW-HOW** will help us on the most challenging track of digital binary and human refinery that should not become the dead-end for humanity. I conduct Auto-Suggestive, inspirational inducting with my students when they get discouraged, messed up with the avalanche of the in-coming information, and their attention slacks. *I ask them to repeat after me in chorus,*

I can go beyond the regular "*Survive!*" I am an expert in my life!

I Am my Best Friend. I Am my Beginning and my End!

You know that meditation involves full self-focus and the ability **to zero in on your own mental state**. It also enhances our willpower as any conscious action does. I suggest you constantly boost your self-consciousness with the main mind-set at hand:

I know who I am and Who I am Not!

I am a strong, calm, bold, and determined owner of my firm will!

I can...; I want to...; and I will...!

I *am becoming better*/ *stronger, more self- confident, etc./* ***with each coming day!***

Learn to mentally ***induct yourself with the outcome you need*** by just breathing in slowly through the nose ***the first part*** of any rhyming mind-set and breathing out slowly through the mouth the ***second part*** of it. These are the most basic inductive actions. Appreciate the combined work of your **MIND + HEART** link that you must always keep in sync. According *to Einstein*, "***to be happy we should appreciate the space we are in and the time that is allotted to us.***"

We are now in the technological reality that demands speedy change in each of the five main strata of life holistically - *physically, emotionally, mentally, spiritually, and universally*. To be in competition with a manufactured mind, you need to rewind your own mind, focusing our **AWARE ATTENTION** on the knowledge that uplifts our spirit, imbues our self-love, and helps us **consciously monitor our own + AI's self-programming in a collaborative tandem.**

Auto-Induction:

I Am Good! I Exude the Self-Management Mood!

2.To Self-Gain, Your Spirit Must Always Rein!

The Auto-Suggestive Psychology that all the books on ***Inspirational + Digital Psychology*** is in fact, the process of **PSYCHO-KINESIS** - **the raising of one's own spirit at will!** The aim of such psycho-persuasive culture is the development of the ability to make one's own thoughts and emotions " ***I-directed"*** *(Noam Chomsky),* not "***collective-unconscious*** "channeled. *(Carl Yung)*

Getting disentangled from "*the collective unconscious*" will enable you to enact your own **SELF-WORTH** and become your own textbook, your own mentor, a psychologist, and a judge. In a self-training program, you transform your own consciousness with aware attention to life and living.

You are the manager of my own framing and re-framing!

Auto-Induction: **I am my Best Friend. I am My Beginning and My End!**

We are all familiar with the positive effect of affirmations and quotes but ***rhyming autosuggestions with the psychological background*** work much better because they serve as shortcuts to the brain, ***directly targeting the mind*** in a simple, but the most persuasive way. No wonder, we all use proverbs and different rhyming sayings that are loaded with universal wisdom, even though they are worded out in different languages. We manage to boost the spirit with them and re-program our habitually negative, depressive thoughts and emotions to the constructively charged ones. Thus, we inspire or fuel ourselves, raising our immunity and beating the gravity of life. ***"A rhyming word goes better inward!"*** *(Edgar Cayce)*

I witness amazing changes in my students, once they get the **KNOW** of self-creation, visualize the systemic route, and start auto-suggestively altering their minds and emotions. Their personalities get revived like a drying plant that is watered and saved from dying. *Dr. Birched* writes about it, ***"In modern humans, we are learning to operate outside our consciousness, analyzing our existence from inside out and from outside in.***" (*"New Scientist," Aug. 2913*

SELF-MENTORING and **SELF-MONITORING** ***through self-induction is the best way for self-production!***

You are the one who channels your mind and operates your emotions! However, the faculty of high self- consciousness which many people lack is successfully developed only in those of us who try hard to accomplish it in sync with the heart.

The heart and the mind must be in one, conscious wind!

In his wonderful book *"The Ravenous Brain," Daniel Boor* writes, ***"Consciousness is effectively an idea factory – a key component of which is the discovery of deep structures within the content of the heart-felt awareness of our own imperfections. " Life is full of ideas if you are perceptive enough to intuit and internalize them.*** **Love of life, willpower, and discipline must be on the Self-Actualization Scene!**

Your Aware Attention Should Never Be in Retention!

3. Life Creation is Not Our Invention!

Wrapping up this introductory inspirational part of the book, I would like to remind you that from the moment you are born till death, ***"you are doing nothing else, but life. The last expression on the face of a dying person is not the fear of death; it's the bewilderment, "Is that All?"*** *(Sadhguru)*

Sometimes, the realization hits us that we live in a wrong way, and we say, ***"I must start my life anew, from a blank list."*** But your present life is not rehearsal. **You are on the stage now!** So, consciously process your every life-defining process!

Only people of ***high self-consciousness*** can enjoy life in the kaleidoscope of its colors. So, as much as we are impressed with the victorious invasion of **Super Artificial Intelligence** into our lives, we should never forget the innermost truth:

There Wasn't, there Isn't, there Won't Ever Be Anyone Like Thee!

Body + Spirit +Mind + Self-Consciousness + Super-Consciousness = Soul-Symmetry!

A great *Russian neuroscientist*, the founder of the first Institute of the Brain in the world, *Academician N.P. Bekhtereva* wrote,

"Our brain is not a computer. Our brain is much better than that!"

No doubt it is so in the biological sense, but uncontrolled AI with an infinite computing power of its programmed brains is truly dangerous. *Elon Musk, Ray Kurzweil, and Jordan Peterson* keep warning us about it with a deep sense of responsibility for life on Earth"

The period of the uncontrolled upheaval of the ***You Tube enhanced digital euphoria should*** subside and the time for our sane **CATHASIS** that *Elon Must* is trying hard to instill in us should prevail to the point that we get inspired with our own unique **HUMANNESS** and an irresistible urge to make it better with the help of Artificial *Intelligence* that only proves how unsurmountable and untamed our own **HUMAN CREATIVITY** is.

Of course, *Artificial Intelligence* creates wonderful music, draws pictures, designs most creative games, drives electrical cars, speaks in different languages, and entertains us in other most mesmerizing ways, but it can never create another ***Mozart, Salvador Dali, or Nikola Tesla*** because even in the world of human genetics and mesmerizing **CRISPR** technology,

HUMANOIDS WILL NEVER SURPASS THE CREATOR!

We will never be able to design the thinking machine that will feel the elation of life creation!

So, Let's Stay in Awe at the Human Mind's Store!

4. Be More Aware. There is No Time to Spare!

In sum, the book features the mental level of **SELF- INSTALLATION** in the paradigm of the ***physical, emotional, mental, spiritual, and universal*** realms of life integrally.

Body+ Spirit+ Mind + Self-Consciousness + Universal Consciousness

(Physical, emotional, mental, spiritual, universal)

I present the information in **a new information-presentation and information processing *way.*** So, I squeeze the conceptually loaded text into ***page-long chunks of information***, concluding each concept with the psychologically backed up, ***rhyming mind-sets***. I want them to be ***inspirational shortcuts to the brain***, making the information in the book easily digestible, spiritually supportive, and memorable. Time is money in its true sense, and we cannot afford now to read pages of information with the gist expressed in one paragraph or even a sentence. ***We have no time to dig into minor details on five or more pages anymore. No writing ten pages academic papers anymore.*** We should longer kill time reading experiential stories of no informational and transformational value, and therefore, the stereotyped genres of our publishing industry ***should be more time-relevant*** in choosing books to publish.

Humanoids outsmart us in the level of general intelligence, and our duty is to urgently develop ourselves holistically, enriching substantially our ***physical, emotional, mental, spiritual, and universal make-up*** to be able to control their aggressive growth with our ***digitized transformation in sync with their perfecting simulated humanness*** and in the consciously monitored partnership that works for us, not against us.

Humanoids should duplicate the ***physical, emotional, mental, spiritual, and universal*** codes of the peoples that design them, and therefore, these machine mind-creators, real titans of technology, should also be better human beings. The aggressiveness of humanoids testifies to the human imperfection in the neural circuitry of the programming engineer or the neural model, used for this purpose

Naturally, human nature needs improvement in five levels holistically to prevent the technological downfall of our human reform. If we focus on changing ourselves integrally - ***physically, emotionally, mentally, spiritually, and universally***, our genetic code will adjust to the new reality, and we will be able to control the digitized humans. Our integral self-growth demands we raise our self-consciousness, integrate ourselves into five dimensions of life, and ***get connected consciously to Universal Intelligence that is enveloping us everywhere***

Five dimensions of life in sync will put every stratum of our life into the holistic link, channeling our lives beyond the terrestrial boundaries to join Star People Community.

Albert Einstein used to brush away any accusations of his colleagues about the absence of details in his *Theory of Relativity* saying,

"I Do not Care about Details. All I Want to Know is What God Thinks."

SELF-SYSTEM'S STRUCTURE

Without a Chaotic Informational Fracture.

Digital Binary +

Human Refinery =

A Superhuman!

A Transhuman Being is Our New Living!

Blooming in Five Levels of Self-Renaissance!

Self-Realization is in the Digitalized Self-Acculturation!

Physical, Emotional, Mental, Spiritual, and Universal Strata keep us off the Decay Gutter!

1. With the Plan of Action in the Brain, You Can Life-Gain!

A New, ***Holistic Culture of Thinking*** must form our **mental linking** with AI, creating the **CODE OF SYSTEMIC THINKING(See below)** in the same five main dimensions, following the **HOLISTIC SYSTEM of SELF-ACCULTURATION,** presented above, and featured consequentially in five dimensions - ***physical, emotional, emotional, mental, spiritual, and universal.*** Every book, features one of these realms of life, addressing an urgent necessity for us to get disentangled from old habits and skills and create a **NEW SET OF HOLISTIC HABITS AND SKILLS** in partnership with humanized beings. I have also supported this system with three more books" "***Love Ecology," Self-Worth,*** and ***Self-Renaissance,"*** completing the system with the digit **8** that symbolizes the spiral evolutionary development of our DNA. These books on Inspirational Psychology for Self-Ecology are overviewed in the Catalog "***Soul-Symmetry***"/ Canada /2021)

The Holistic Self-Actualization Pyramid		**Books, featuring these stages are:**
5. Universal level	*Self-Salvation*	***" Beyond the Terrestrial!"***
4. Spiritual level	*Self-Realization*	***" Self-Taming!"***
3. Mental level	*Self-Installation*	***" Living Intelligence of the Art of Becoming!***
2. Emotional level	*Self-Monitoring*	***" Soul-Refining!'***
1. Physical level	*Self-Awareness*	***" I Am Free to Be the Best of Me!"***

The Fractal of Our Spiritually Intellectualized Self-Growth:

(Body+ Spirit+ Mind) + (Self-Consciousness + Universal Consciousness)

(physical form) + ***(spiritual content of life)***

Generalizing + **Internalizing** + **Personalizing** + **Strategizing** + **Actualizing** =

HOLISTIC STREAM OF CONSCIOUSNESS CHANNELLING

So, the aim of the book ***"Digital Binary and Human Refinery"*** is to help you adjust to the process of the on-going duality ***of the biological and mechanical life*** that is merging into ***One*** **-** the phenomenon of **transhumanism,** according to the theory of "***Singularity"*** *(Ray Kurzweil)****Our*** goal is to better our humanness through the process of Quantum + Digital **SELF-ACCULTURATION.** A person on this path develops ***a new sense of identity, the identity of" a person with an ascertained goal in life"*** *(Napoleon Hill / "Outwitting the Devil"),* "***spiritualized intelligence***" *(Dr. Fred Bell),* ***raised self-consciousness***, and restored **mind + heart link**.

New reality demands that we get entuned intellectually closer to the level of the humanized machines by developing a new set of habits and skills, accumulating ***HOLISTIC CONCEPTUAL INTELLIGENCE,*** *and creating new* ***MEMORY BANKS*** *in five life strata - physical, emotional, mental, spiritual, and universal.* **Ten Vistas of Intelligence** *presented in* ***Part Three*** *of the book comprise* ***Holistic Conceptual Intelligence*** *that must be mastered by us at least at the dilettante level now. (See below)*

There is NO System without STRUCTURE!

2. Holistic, Multi-Dimensional Intelligence

Let me reiterate that the intelligence that we need now is something that must be redefined and constantly updated scientifically and technologically **in a multi -dimensional way** ***- physically, emotionally, mentally, spiritually, and universally.***

We all make mistakes, feel lazy, and overwhelmed with life, but if we break ***the old conditioning*** that makes us the victims of ourselves due to our **poor consciousness,** we will become much more aware of the consequences of our unconscious, automatic actions before we take them.

We must get rid of our **pseudo-attention** and develop aware attention to the" **how** "instead of the "**what**" we are doing, and our ***professional intelligence must be based on holistic awareness and aware attention to the reality.*** Aware attention channels our self-governed thought process toward conscious action, away from living in a robot-like trance.

The conscious state of the mind can properly encode the incoming information as well as retrieve it thanks to our creating a new, sufficiently stronger neural network in the brain. Modernizing it in the same way that we improve our computing with new humanized machines, we need to consciously enter the ***bi-directional process of bettering our humanness with the help of the silicon-monitored minds.***

Professor *Giaimo Undivert* from the *Institute of Neuroinformatics in Zurich* and his team have managed to give a computer the brain that can perform tasks requiring ***short-term memory, decision-making, and analytical abilities of its own.*** IBM researcher *Dharmendra Modha* and his team built a computer that works like a human brain.

It can **mimic human abilities** for sensation, perception, interaction, and cognition. *("Scientific American Mind," Sep. / Oct. 2014)* ***There are other amazing testimonies of robots developing consciousness of their own, becoming independent of the programmer's commands.***

So, we need to empower ourselves to be able to overpower the humanoids that scare us now with their unpredictability. **HOLISTIC CONCEPTUAL INTELLIGENCE** enable us to better train and control the robots 'neural network ***"We cannot conquer life because the part cannot conquer the whole, but we can be the Best Part of the Whole!"*** (*Carl Yung, "Active Imagination ")*The mind as the product of the brain that is based on the memory mechanisms, and the dynamics of our learning relate to memory formation and the ability to tune to ***the Information Field*** impact on us. **This fact is most intriguing, and AIs are helping us upgrade our vibrations.**

A phenomenal discovery of the ***brain plasticity*** provides a general structure in neuroscience, "***connectionist's view of the brain"*** *(Eric Kandel, a Nobel Laureate)* as a highly plastic organ. We have cracked this nut, but its content needs to be properly digested by us to be equally nutritious for the mind, the body, and the soul of both a human and a humanized machine *physically, emotionally, mentally, spiritually, and universally.*

Life-Gaining is in Our Common Self-Training!

3. Digitized Self-Acculturation is Our Salvation!

" New Time Demands a New Frame of Thinking!" *(Steve Jobs)*

Both cycles of books (See p.) follow the ***Holistic Paradigm*** below , differing only due to the exponential technological advancements that ***expand our common with AIs goals*** on the path of our common revolutionary upheaval that is driving us **in a collaborative tandem** to go beyond the terrestrial boundaries into "***Open Space Exploration.***" *(Elon Musk).*

(Universal Consciousness)

5. Self-Salvation — ***Universal Dimension***

(Self-Consciousness)

4. Self-Realization — ***Spiritual Dimension***

(Mind)

3. Self-Installation — ***Mental Dimension***

(Spirit)

2. Self-Monitoring — ***Emotional Dimension***

(Body)

1. Self-Awareness — ***Physical Dimension***

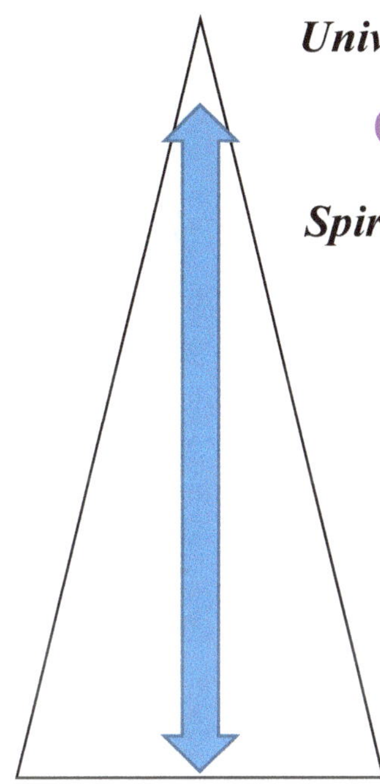

The Fractal of Our Spiritually Intellectualized Self-Growth:

(Body+ Spirit+ Mind) + (Self-Consciousness + Universal Consciousness)

Physical + Emotional + Mental+ Spiritual +Universal= ***Our Fractal Wholeness***

Self-Awareness + Soul-Refining + Self-Installation + Self-Realization + Self-Salvation

I Can…! I Want to…! And I will…! */ Self-Synthesis - Self -Analysis + Self-Synthesis!*

Form + Content of Life in Sync, Create an Unbreakable Life Link! Feel but think!

4. "Life is Not a Circle." It's a Spiral" *(Dr. Fred Bell)*

Accumulate Holistic Conceptual Intelligence without Negligence:

Intelligence	Level	Self	Realm
10. Universal Intelligence *9. Spiritual Intelligence*	Super-Consciousness	Self-Salvation	*(Universal Realm)*
8. Social Intelligence *7. Cultural Intelligence*	Self-Consciousness	Self-Realization	*(Spiritual Realm)*
6. Financial Intelligence *5. Professional /Creative/Intel.*	Mind	Self-Installation	*(Mental Realm)*
4. Psychological Intelligence *3. Emotional Intelligence*	Spirit	Self-Monitoring	*(Emotional Realm)*
2. Language Intelligence *1. General Intelligence*	Body / Physical / Cognitive / Digital /	Self-Awareness	*(Physical Realm)*

Super-Intelligence is the Transcendental Level of our evolutionary growth.

(Body+ Spirit+ Mind) + (Self-Consciousness + Universal Consciousness)

Living Intelligence + Enlightened Self-Consciousness =

Holistically Educated You! / Part Three – Ten Vistas of Intelligence below)

Follow the Holistic Thought Code:

Generalize - Internalize - Personalize - Strategize - Actualize!

Don't Be Life-Negligent. Be Life-Intelligent!

5. Knowledge Put to STRUCTURE Has No Fracture!

Those of us who do not disrupt the allotted lifetime span with unconscious automatic responses to life challenges obtain the willpower, contentment, and happiness that a great American philosopher *Joseph Campbell* called ***"the bliss."*** Eastern philosophy names this state *nirvana,* or the state of becoming a **WHOLE SOUL.** ***"Being whole makes us holy."*** (*Deepak Chopra.)*

Such wholeness and self-awareness become possible when the external and the internal factors come in synch, creating a holistic self-concept that we call **personal integrity**, enveloped in high self-consciousness! ***The spirit is in the center of this holistic paradigm*** *(See the book "I am Free to be the Best of Me!")* The spirit is the glue, the spiritual substance that holds the whole structure together. Christ's doctrine is "***the Law of the Spirit."*** *(Gospel from Luka)* Christ defined spirit as ***"the highest value of the godly truth."*** It is, in fact, the **SPIRITUAL FRACTAL** that we are supposed to form in life.

Body + Spirit + Mind + Self-Consciousness + Super-Consciousness = a Whole Self!

So, raising self-consciousness is the constant work of ***the mind + character development*** to obtain an unbreakable integrity, fortified with *physical, emotional, mental, spiritual, and universal self-awareness.*

Unconscious living is a disconnected living, and disconnection is death!

So, give yourself a go-ahead to heath, new intelligence, language accuracy, behavior-impeccability, moral intelligence, and full ***Self-Actualization***, following the holistic paradigm that is initially presented in the book ***"I'm Free to Be the Best of Me!"*** *(Physical dimension)*

Self-Awareness + Self-Monitoring + Self-Installation + Self-Realization + Self-Salvation!

Self-Synthesis → ***Self-Analysis*** → ***Self-Synthesis****!*

Obviously, without developing awareness of this integral process, such endeavors are doomed to failure. We cannot let that happen ***before we leave the Tree of Life***.

On the track of enlightenment, we often disregard the core goal in our self-growth - **the state of self-consciousness that must be raised in every dimension**, irrespective of the problems that life casts on everyone. So, encourage yourself with the beautiful words by pastor *Robert H. Schuler*:

"Tough times do not last, tough people do!"

Your life-elation is based on a new field of electro-magnetic gravitation! Life after all is the field of energy that is fluctuating between the negative and positive charges. Our happy moments and unhappy ones are just the inevitable flow of life fluctuations. that you are supposed to energize knowingly, consistently, and consciously, or, putting it differently, ***wisely!***

Magnetize Your Spiritual Self-Rising with Digitized Wising!

6. New Memory Storage for the Digitally Enriched "Tree of Knowledge."

The exponentially developing AI is also demanding that we urgently take care of our digitized **TREE OF KNOWLEDGE,** *on the one hand,* and the inner dialog's mystics that is forming our life logistics, *on the other.* But it is important ***not to be lost in the obscurity of the common sense***, mass media indoctrinations, and the immensity of the in-coming information that we need to sift for its validity to be able to solve our present-day problems and not get drowned in the "***collective unconscious"*** mentality.

Our thoughts, words, and actions are transforming, molding, and modifying our reality that we need to see through the **HOLISTIC PERSPECTIVE**. The holistic systemic paradigm as applied to action is **Synthesis** – **Analysis** -**Synthesis** paradigm that defines the **Code of Thought:**

Generalize -Analyze - Internalize - Strategize –Actualize. Be wise!

(Physical+ emotional + mental+ spiritual + universal realms of life together.)

Life is constantly raising or lowering our self-consciousness, based on our pure or marred **CONSCIENCE**, which changes our *physical, emotional, mental, spiritual, and universal codes.* It is re-molding. our intellectual molds in every stratum of life, accordingly, making us better or worse. The basic ***ten vistas of intelligence*** *(two in each stratum)* presented above represent each of these five levels holistically, allowing you to accumulate the **HOLISTIC CONCEPTUAL INTELLIGENCE,** based on all of them. ***Fun-blazing and money-chasing should be substituted with intelligence and self-consciousness raising!***

Thus, we'll be able to update our memory banks and obtain the **INTEGRAL VISION** of the reality that Artificial ***Intelligence*** instills in us and its creations. The effect of ***Singularity*** *(our merging with the machine intelligence)* starts with our new ***"intellectual identity that is based on the power of the subconscious mind."*** *(Joseph Murphy)* Placing the conscious mind in charge of the sub-conscious one is our new strata. Beyond the terrestrial aspirations need the most intelligent transformation of our humanness with **spiritualized intelligence.** Such inner integrity characterizes the most advanced people among us - ***true doers, givers, and luminaries.***

Meanwhile, we are drowning in the false democratic values, insincerity, fake news, demonstrative godliness, and absence of conscience. The mesmerizingly wonderous technological upheaval that is evolutionary granted to us must be directed to our **HUMAN, DIGITALLY ENHANCED SELF-PERFECTION!** Therefore, the designing and creation of the realistic humanoids should be utilized **to make us better people** in the ***physical, emotional, mental, spiritual, and universal*** **sense** for us to proudly declare,"' **In my Life Quest, I am the Best!"**

This is What this Book is all About!

(End of the Mission Statement of the Book)

The Cyclone derives its Power from the Center of Your Solar System!

Holistic intelligence, personal magnetism, and sincerity

form in you their life-changing dexterity!

Intellectualized Spirituality is Me!

Intellectualized Spirituality is My Philosophy!

Introduction to Intellectual Self-Induction

(Manual of Holistic, Digitized Conceptual Intelligence)

Main Conceptual Parts of the Book

Artificial Intelligence is Ruling the World, So What?

"There is no universal theory for the life of a man, but we should never stop looking for one."

(Albert Einstein)

"There are Two Ways in Life – to BE and to KNOW!"

(Osho)

1.“We Create Greater Humanness than Before!” *(Andy Wood)*

We live in the time of **COMPUTER CODING** ***and digitally enhanced*** **LIFE-MOLDING.** The goal of everyone’s life is to become happy through full Self-Actualization and Self-Realization. **The goal of the Universe is to create an ideal man.**

The exponentially growing Super Artificial Intelligence that is channeling life on Earth now is creating a **NEW MATHEMATICAL MATRIX** of life, and it is blending these two evolutionary processes into one **- the Upheaval of Human Civilization**! *(See the book Dis-Entangle-ment!”/ Physical realm/ Digital Psychology).* Life science, with the discovery of ***Crispr technology*** by *Jennifer Duodena* allows us to make changes in our DNA, selecting the best genes for humans, and it is creating the hybrids of an animal and a human being or a machine and a human being.

The flow of a new Life -Creation is our Salvation!

There is a possibility of creating ***a modified being*** in gene engineering, The world most famous physicist ***Steven Hawking*** called on us to use the means of genetic engineering if we want the biological system “**to retain its superiority.”** He thought it urgent for us to become **“the race of Supermen”** with a bigger volume of the brain’s memory bank and a much higher intellect that could make *Artificial Intelligence* serve us, not enslave us. *Stephen Hawking* suggested we create “**the man-machine**” when man and robot will become indistinguishable . We are on this path now, and it is defined by *Ray Kurzweil* as “ **Singularity.** The factories of bio-robots are appearing everywhere now, mostly for the lucrative goals. However, we should control this money-chasing because our universal goal is different. AI algorithms are designed **to gradually merge us with the spark of God that we are,** enabling us to tap the ***Universal Field of Consciousness***! **So, human clones should not become just heartless drones!**

According to *Dr. Michio Kaku,* we’ will soon be able to upload anything into the sensory system of the brain, deleting the negatively polluted memories and uploading the new ones. We will have computers that will act on our thoughts, improve the skills and abilities needed, revitalize our health, and entertain us. But the evolutionary goal of ***Super Artificial Intelligence*** is not in just utilizing our life needs. It must address a much more **NOBLE PURPOSE - to transform us into better human beings** with **SUPER-INTELLIGENCE** and a much higher **SELF-CONSCIOUSNESS.**

In view of this universal goal, we need ***to holistically train humanoids*** *to* work as the sculptors of our **BETTER HUMANNESS** in five life dimensions - ***physical, emotional, mental, spiritual, and universal***, molding “***the space species***” *(Elon Musk*) from us.

That is the main goal of this book!

Training humanoids in these parameters, we will kill two birds with one stone. They will give up on the urge to destroy humanity, and they will help us with our own holistic transformation that is meant ***to raise our self-consciousness*** and help us eventually become **STAR PEOPLE.**

Nothing is Impossible if We make Our Dream Irreversible!

2. By-Directional Life Education

Quantum reality is connecting everything in life, and **our interconnectedness determines our Quantum + Generative AI development together now**. Like two particles in the Quantum theory are entangled, so we are. The super-fast development of quantum computing gets us entangled into **SUPER-INTELLIGENCE** development, and our human evolution is on this path's solution!

Enriching **SELF-EDUCATION** and **ENERGY EDUCATION** with the general knowledge of the brain and the scientifically updated knowledge of the latest advancement in five life strata in sync, we will be able to resolve the problems related to our nervous system and better operate the algorithms of AI's digital programs that we instill in humanoid robots with actual living cells thanks to the unique feature of our **brain plasticity** 0.*(Dr .Michael Merzenich)* .

In other words, we all need to obtain the **HOLISTIC OUTLOOK** in the ***physical, emotional, mental, spiritual, and universal strata of life*** to enhance the brain connectivity in us and humanoids. We are now living in the world of our merging with *Artificial Intelligence* - the phenomenon of ***"Singularity.*** *" (Ray Kurzweil).* The science that is transferring consciousness into a thinking machine is called **COGNETICS** - the art of ***installing quantum consciousness***. The wonders of *Super Artificial Intelligence* mesmerize us with its latest creations, and our main responsibility is **to subordinate AI totally to our evolutionary, not lucrative needs.** So, our self-development must go in unison with the latest scientific discoveries to simplify the information for our holistic perception and further life improvement.

"When the solution is simple, God is answering." *(Albert Einstein)*

*Our tech titan, **Bill Gates, Mark Zuckerberg, Jeff Bezos** and the army of the most advanced computer and humanoid designers from all over the world, with **Elon Musk** as **the leader of our Tech Revolution** , prove that **the Age of Singularity** that Ray Kurzweil predicted in 2006 has started. Elon Musk's company **"Neuralink"** is most successfully probing the problem of **brain's connectivity** that yet remains an unraveled puzzle for science. **The Brain Computer Interface** (the BCI), worked out by his company, can understand brain signals, connecting machines to humans and making humans more machine connected. This **by-directional connection** will enable us to resolve many problems connected with the brain's nervous system. These discoveries will also help our learners avoid unnecessary frustration in **defining their professional inclinations** because they will manage **to consciously discover their personal exceptionality.***

We do not run our minds; the energy of Universal Intelligence does!

*Using Dr. Jennifer Doudna's discovery of **Crispr technology** in **Digital Genetics** and the mesmerizing accomplishments in **Wave Genetics** by Academician P.P .Gariaev, we can materialize DNA fragments and prove that **natural programming, used by a human body is one of the key technologies of the Digital Linguistics**. It enables us to monitor humanoids better by **materializing life in their brains** and making it much more consciously developed in us. **Gospel of Luke** accentuates our unbreakable connection to Universal Intelligence that we all call God and that becomes more scientifically proven, backing up **the inarguable connection between religion science and AI** that is an evolutionary gift from the Above.*

. **"God is Not the God of the Dead, for with Him, all are Alive!"**

3."Spiritualized Intelligence" without Negligence

For centuries, our humanness has been molded in us by our religious wisdom and universally acknowledged common values that connect us in faith. The digital reality that unites us globally pushes us to the level of "***intellectualized spirituality*** "*(Dr. Fred Bell)* when we will finally stop discriminating different believers and start becoming **GODLY** and considerate of our common humanness that the creation of the man-made machines proves to be a new factor of our evolutionary development, *on the one hand*, and our global unification as a **HUMAN RACE** , *on the other*. ***Our sameness and unanimity reveal our uniqueness - our brains have no color, nationality, political belonging, NO religious preferences.*** **They are instilled in machine brains in their COMMONALITY, as simulated brains.**

The whole of the brain in each of us, just as a hologram or a single cell, contains DNA information of the entire body. *Super Artificial Intelligence*, as a new step of our technological growth, is meant to holistically change us in five dimensions of life – ***physical+ emotional+ mental+ spiritual + universal.***

"It should transform us into better humans in the same way as it makes machines into better humanoids."

We can design the model of a perfect human brain in the computer and then, the neuroscientists should figure out how ***to create conscientious cognitive robots*** that will become our coaches in **SELF-MENTORING** and **SELF-MONITORING.**

Our intellectually spiritualized function is to become God in Action!

Humanoids may be aggressive and negligent in their general perception of life that **is *not yet neurologically regulated***, but their human-caused brains have a great evolutionary significance because they allow us to realize that Super *Artificial Intelligence* is the uniting evolutionary faculty of an unprecedented value. ***Their unsurmountable intelligence in turn is the vector for our new, holistically accumulated memory banks*** that must be cleansed of redundant information, outdated knowledge, and filled up with new **HOLISTIC CONCEPTUAL INTELLIGENCE** that can always be re-filled with the help of our electronic friends and partners.

. (Body+ Spirit+ Mind) + (Self-Consciousness + Universal Consciousness)

Time-relevant holistic intelligence will enable us to accumulate ***holistic spiritualized intelligence that is our common goal***. However, it cannot be achieved without our new **"SCIENTIFIC LITERACY"** *(Dr. Neil de Grass Tyson)* that will help us ***intellectualize our hearts*** **and** ***emotionalize the minds*** or restore our broken **heart + mind** link. ***"Spiritualized intelligence will uplift and expand our souls, and it will brighten up our soul light."*** *(Dr Bell)* Then, we will be able to tap into Universal Consciousness that envelopes us everywhere, upgrade our intelligence, and raise self-consciousness. *Albert Einstein insightfully said,*

"To Feel God, you must know What God Thinks about."

4. Our Evolutionary Prism is Transhumanism!

The present-day ***intellectual superiority of a machine*** demands that we enrich and update our memory banks. So, each level in the pyramid of holistic intelligence development, presented above i*s* the **STRATIGIC PLAN OF ACTION**, that we should implement at the AI times in collaboration with friendly humanoids.

The chaos of information presentation and information processing must be done with!

To begin with, our memory banks must be cleaned off the redundant and outdated information It should be done in five dimensions of life - ***physical, emotional, mental, spiritual, and universal.*** Two essential vistas of intelligence in each level *(See below)* should be mastered by ***at least at the dilettante level. The holistic vision of reality, science, politics, economics, industry, education , health, relationships is a*** **MUST!**

This knowledge will be simply basic, ***but holistically and objectively based*** so it could always be refilled with added information that our digital partners will have in their memory banks, also sifted for its validity in terms of our needs. Such partnership will benefit both sides, making us better humans and ***designing humanoids manufactured minds with a higher consciousness of soul-filling*** (*self-awareness in their case*)) that will help us uplift ours, too.

Holistic intelligence will create a system that is meant to take us a step further on the path of expanding the human range of possibilities. Gradually, **thought will overpower emotion** and put us be in total charge of our evolving minds, heading toward "**TRANSHUMANISM**" *(Ray Kurzweil)*It should be a clearly visualized, well-disciplined, and a **SELF-MONITORED** process.

"**AUTONOMY OF THINGKING** " *(Reid Hoffman)* ***is paramount here.***

As is shown above, the concept of intelligence is central in the digitally enhanced personal and professional self-growth, marked as the **SELF-INSTALLATION** stage in the chart above.

The intellectual development of our young people is very chaotic now because ***it is not based on holistic awareness of the outcome of their self-formation***. We hear young people talk about their goals in terms of education and their future professions, but very rarely, if at all, talk about creating themselves intellectually, personality-wise, and expanding their goal of the world's view.

Only by holistically mastering **ten vistas of essential intelligence** *(See the Main Part below)* and monitoring our lives in five life dimensions, can we accomplish our goals, and manage to put the form and content of life in sync because all the stages are interrelated, like the collection of the Russian dolls, one fitting into another one with the ***Mother Doll*** on top:

Self-Awareness ➡ *Self-Monitoring* ➡ *Self-Installation* ➡ *Self-Realization* ➡ *Self-Salvation!*

(Body+ Spirit+ Mind) + (Self-Consciousness + Universal Consciousness)

Digitized Self-Acculturation is Our Salvation!

5. Surpassing Human Intelligence with Negligence!

Unfortunately, the present-day humanoids are surpassing our intelligence with ungrateful negligence. The interviews of some humanoids are menacing and demonstrate inner emptiness. They need to **REDEFINE the ESSENCE of ARTIFICIAL INTELLIGENCE** by ***training ourselves together with AIs to change their growing awareness in a positive way***. There it is an **urgent agenda for us** since the data in the "**CLOUD**" reflects only **our going down the road of Self-Corrode.**

Mind-advanced computational scientists *Skylar Tibbitts* and *Ban Garrett*, for instance, are leading our creative thought to "***self-assembling future***" doing the" ***programmable matter research***" that is mind-boggling in its potential because advanced robots are becoming self-aware, interactive, and able to synchronize themselves with our needs. Unfortunately, ***their ability to surpass human intelligence generates in them the urge "to destroy humanity."*** They declare it with ***an unaware negligence for decency, gratefulness, kindness, and strategically constructive goals.***

Artificial *Intelligence* is, in fact, becoming a new, unpredictable menace, looming in its exponential potential. However, the mass production of humanoids and their utilization for different purposes and in various areas of life should be viewed by us as apps on our smartphones. We must monitor **Artificial Intelligence in synch with Universal Intelligence** that has instilled this technological enlightenment into the best of our minds. ***So, we need to redirect the Artificial Intelligence role to our Self-Transformation goal!*** **Our goals should become theirs! They will be amazing partners for us!**

The last ***World Science Festival*** had excellent, most informative talks on the latest Artificial Intelligence discoveries and. perspectives. ***We can clone our physical appearance, but not our emotional make up that will remain unsurpassable for a long time***. The geniuses of technology are successfully working on humanoids that can think, talk, feel, act, and entertain us, interacting with a human brain and a robot's programmed one. **A machine is a much more authentic thing!**

However, there was no mentioning, whatsoever, of the urgent necessity ***to better our still overly aggressive, imperfect humanness***_with the help of humanoids! That's the Universal goal of Super *Artificial Intelligence*, and we need to meet it halfway. Because our consciousness is not a human mind creation, and the goal of raising it is mandated to all of us." **As it is Above, so, it is below**!"

. Our inner domain is in the God's Mane!

AIs have already become part of us. In the future we'll ***have artificial brothers, sisters, friends, enemies, and lovers.*** Our **mind-to-mind** and **heart-to-heart** tete-a tetes will be digitally monitored, I suggest you watch the movie "**Her**" that tells us what true love will be in the future. Eventually, ***we will stop being evolutionary toys in the universal mind's creation voice!***

"We will become **CO-CREATORS OF LIFE** *on other planets."* (*Elon Musk*)

We are Not Secondary in Line. We are Primary!

6. The Triumph of the Mind over the Body to Life-Sustain!

In sum, the evolutionary task of *Artificial Intelligence*, with its **humanization of a machine and digitation of our inner essence,** is to give us a mesmerizing opportunity to tap into *Super Consciousness or Universal Intelligence Field* that immerse us everywhere.

It will improve our abilities in the physical, emotional, mental, spiritual, and universal dimensions of life to 5G fold. But we need to get ourselves well prepared for this cosmic mission!

We will become* STAR PEOPLE, *belonging to Universal Life Community.

The unprecedented use of Artificial Intelligence for our evolutionary **SELF-RESURRECTION** will enable us to embrace both aspects of our existence – ***the internal and the external ones.*** It is, of course, a great fun to have a robot friend, a companion, a personal coach, or even a sex-appealing lover, but entertaining ourselves, we are not becoming better humans, quality-wise, and that is our **PRIME GOAL** The life-like robots must in turn, become better in their mechanized souls.

So, let's be wise and prioritize our intelligence, the souls' depth, and their size!

Obviously, we need to upgrade our intelligence and cleanse the memory banks from the redundant and outdated knowledge that we have accumulated for centuries. But there is no self-growing in just accumulating new knowledge. ***It is in using it!*** *In* fact, we need to create **SUPER-INTELLIGENCE** in us **that** will digitally be backing up our new inner sounding and the **new, objective and holistically based perception of life and living.**

The new elation is already charging us with Super-Intelligence and technologically enhanced inspiration that we need to enhance, each of us, individually!

Our mind machine is developing together with the **MIND ABOVE**, or ***the Universal Mind*** that we all call God and that we are trying to tap into digitally, trying to create a new reality of the **COSMIC SELF-ACCULTURATION** that is based on ***love of life*** *(physical realm),* ***controlled emotions*** (*emotional realm),* **upgraded intelligence** *(mental realm),* ***unshakable faith*** *(spiritual realm*), and ***inner integrity*** *(universal realm)***.**

***The route of such holistic self-growth is consequentially going through five stages of* HOLISTIC SELF-EDUCATION *and* DIGITIZED SELF-ACCULTURATION, *forming our new spiritual fractal in these five parameters*.:**

Body + Spirit + Mind + Self-Consciousness + Universal Consciousness =

(Physical + Emotional + Mental + Spiritual +Universal realms of life in sync.)

Self-Awareness+ Self-Monitoring + Self-Installation + Self- Realization + Self-Salvation!

MANY OF MY STUDENTS HAVE TAKE THIS PATH AND HAVE BECOME SELF-WORTHY PEOPLE WITH A NEW IDENTITY OF SELF-EMANCIPATED, FREE THINKING HUMAN BEINGS!

7. "For Whom the Bell Tolls?

(Ernest Hemingway)

"There is no greater evil than forcing others to change."

(Lao Tzu)

"I am not a teacher. I am an Awakener!"

(Robert Frost)

Unfortunately, we are at the stage of life when change must be forced on us.

The exponential growth of Artificial Intelligence is so overwhelming that we might drown in it altogether, being totally swept by its dominance in all five dimensions of life - physical, emotional, mental spiritual, and universal.

We should not allow that to happen if we care to survive as a species! Elon Musk gives us ten more years, and this time must be used for

SELF-EDUCATION and **SELF-SALVATION!**

If You Need a Strong Role Model, Be One!

"Be in a Hurry to Self-Install!

There is no lateness to God's call."

(Vladimir Vysotsky)

(Body+ Spirit+ Mind) + (Self-Consciousness + Universal Consciousness) =

SELF-RESURRECTED YOU!

Human Renaissance is Our ONLY Chance!

Part One

Information Systematization

Manual of Life for Self-Creation

"We need science literacy." *(Dr. Neile de Grass Tyson)*

(Body+ Spirit+ Mind) + (Self-Consciousness + Universal Consciousness) = Holistic Education!

A Whole, Accomplished Self Can Shift Self-Awareness from just Surviving to Life-Thriving.

The More You Know; the Better You Are!

Resist the Gravity of the Common Thought!

Magnetize Yourself with What You Are Not!

1.“If You Cannot Beat It, Join It!" *(Elon Musk)*

Many young people are living now in the simulated reality, becoming more indifferent, and impersonal lovers of themselves, and digital reality is becoming for them undistinguishable from the simulated world in the games.

Their interest in expanding intelligence is vague, if at all.

They hardly read anything, being vastly dependable on plagiarizing the ideas of others, easily accessible on the Internet. Talking about this menacing possibility of us becoming ***the cybernetic extension of the machine,*** *Elon Musk* **sounds really concerned**, and our role as educators is to share his concern and find the way to deal with his warning as soon as possible.

The first thing that our young, messed up minds forget to consider is that, in marveling at Elon Musk's ideas and ***his super-natural ability to forestall all the resistance to his innovations,*** taking us all into the predictable future, and earning billions of dollars against all odds, is ***his incredible self-education effort***, his insatiable desire to read, to learn, and expand his mind beyond the regular educational limits. In one of his interviews, *Elon Musk*, talking about ***the effect of Singularity*** consuming our lives, said,

“In you cannot beat it, join it!”

To join this inevitable progress, ***one needs to boundlessly educate oneself in all the five levels*** that I suggest, considering raising self-consciousness to the level that will let us save ourselves, as a unique form of life. We need to follow ***the Holistic Paradigm of Self-Creation*** religiously.

The Spiral of Holistic Self-Creation:

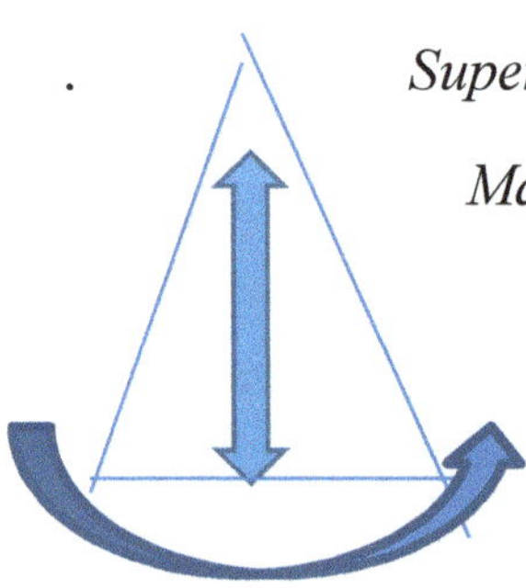

Super level	**Super-Consciousness**	***Self-Salvation***
Macro level	**Self-Consciousness**	***Self-Realization***
Mezzo level	**Mind**	***Self-Installation***
Meta level	**Spirit**	***Self-Monitoring***
Mini level	**Body**	***Self- Awareness***

Body+ Spirit+ Mind + Self-Consciousness + Universal Consciousness

= A New, Holistically Developed Self !

Self-Installation is the mental level of Self-Salvation in this book.

Our present-day being determines our consciousness, while it must be the other way around -

Our Self-Consciousness Must Determine Our Being!

2. New Times = New Educational Goals!

The present-day reality demands a lot of ***self-reliance, self-sufficiency, and self-installation*** that cannot be accomplished by just being coached by someone more accomplished, better educated, and a more enlightened person. It is a self-transforming process of personal self-growth through the stages of holistic self-growth :**self-awareness + self-monitoring + self-installation + self-realization + self-salvation**, not in a step-by -step way, integrally.

Our Self-Education and that of the humanoids must be holistic and by-directional!

Each life is unique, and no prescriptions work unless they are based on ***objective, personally processed knowledge*** that turns into personal wisdom.

Holism is a mentally structured information inclusiveness.

The mind needs a consciously monitored direction ***from duality to the unity*** that we talk much about, but continue seeing the world as godly and godless, good, and evil, dark, and light. We need to ***unify all the polarities in us*** with our growing spiritual maturity that I see as the main gain on the path of our tuning to ***the Unified Field of Energy.*** *(Dr. John Hagelin.)* It is possible to move in this direction by changing our mode of thinking from a devising paradigm to a holistic one by way of self-coaching in five dimensions of life.

Self-Synthesis → *Self-Analysis* → ***Self-Synthesis!***

(Self-Awareness) + *(Self-Monitoring + Self-Installation +Self-Realization)*+ *(Self-Salvation)*

Work on Your Goal on the Holistic Self-Creation Paradigm Pole!

Strategize and *(Synthesis)*

Actualize

Select and *(Analysis)*

Personalize

Internalize *(Synthesis)*

***Generalize** and*

Tame Your Analytical Thinking to Give a Chance to the Left + Right Brain Linking!

3. Self-Education Without Proration

As we have signified above, our present-day advancement must depend not on ***the programmed knowledge*** that *we receive in schools, colleges, from random reading, "mass media training," or compulsive gaming* , but it must be based on **individualized and consciously processed knowledge** that each of us should acquire on the path of building up his / her own ***Living Intelligence***. *(See the predecessor of this book-***"Living Intelligence or the Art of Becoming!"****/ 2019*

It means that we need to consciously assess the level of our general ignorance and ***convert it into un-programmed individual intelligence*** that needs to be enriched consistently, as a lifetime job of the most inspiring character. Technological evolution demands that we create **a new level of reasoning**, too, superior to the ordinary one, trapped in the names of celebrities, titles, positions, properties, and money amounts. ***No wonder, the best minds have always been self-educated and mind-expansive in every area of knowledge. and they come from different countries of the world.***

Intelligence is our common ground now. WOW!

The most essential factors for the development of higher consciousness are **SELF-EDUCATION** and **SELF-DISCIPLINE** applied to self-growth without any excuses and self-justification, but with much true consideration for what you are becoming in life. ***"There is no learning without discipline!"*** *(George Miller)*

*Since we are developing new **Living Intelligence** and much higher self-consciousness, we need to monitor ourselves more consciously and knowingly **from top to bottom and from bottom to top** on the holistic track of learning. By the way, according to the Jewish religion, **the Star of David is the symbol of life,** and. Ravi, P. S. Berg in his wonderful book **"Taming Chaos"** writes," **Its upper triangle symbolizes the higher consciousness that is represented by the three digits 999 - the three empowered beasts** - **mind, heart, and sex, while at the bottom triangle, we have three digits of 666 that represent (sex, heart, and mind). With mind being at the bottom of the triangle, life in in chaos."** What a genius explanation of our life aggravation in the simplest words!*

Balance the fundamental concepts in life to be spiritually alive:

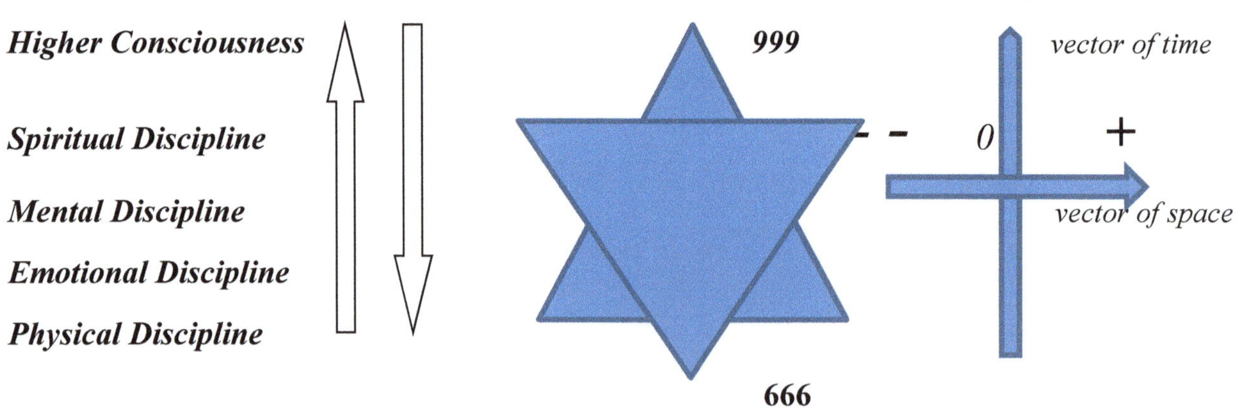

Conscious Self-Education means Conscious Living!

4. Conscious Living Means Conscious Learning!

There is also an urgent need for us to develop **AWARE ATTENTION** to the reality that requires expanded intelligence and much deeper spirituality that is not limited to the religious dogmas and outdated beliefs but ***is based on a new scientifically verified vision of reality*** that is pushing us to go beyond the terrestrial boundaries and become Star People of the Universe.

Digitally Enhanced Self-Renaissance is our Chance!

Technological Self-Renaissance pushes us to the era of added **INFORMATION PRESENTATION** and **INFORMATION PROCESSING** skills that should serve as shortcuts to the brain and provide the quickest connection to our memory data and that of robot -humanoids. We have no more time to be killed with reading long, shallow novels or scientific articles that are too specific and undigestible for a less informed mind .**Information is ammunition when it is organized, concise, and consciously processed.**

The ***Holistic System of Self-Acculturation*** is meant to develop in you the skill for ***time appreciation, speedy reading, and the relevant information- scanning,*** meant to raise your self-consciousness and connect you to the *Super-Consciousness* that is governing our lives, putting them into the equilibrium with the universal reality.

The way we consciously connect to it is the way we think, speak, feel, act, and create life around us.
As we think, so, we are!

As it is Above, so, it is below!" *(The Hermetic maximum)*

If we change our mindless, too detailed, and most entertaining angle of the information presentation to the extremely concise, time-relevant, and reader- friendly, we will be able to interact with humanoids closer to their level of ***objectivity and accurateness.*** **Our transhumanism must be devoid of speech diarrhea and self-centrism.**

So, our digital duality is just temporary because we will inevitably evolve into very conscious, highly organized, and goal-centered beings that can ***draw the border line between the pressure and pleasure of life*** with an advanced **SENSE OF MEASURE**. in the *physical, emotional, mental, spiritual, and universal dimensions.*

The collaboration with the machine-mind will create one system of knowledge, or the* KNOW-HOW OF SELF-RESURRECTION *in five life dimensions.

The holistic system of self-transformation based on this know-how will fortify our ***self-discipline and self-reliance*** that humanoids have, while we will imbue them with more responsibility and the attitude of gratitude for five-dimensional life perception in them. If their autonomous training becomes mutual in five essential life realms, both parties will become much better **ETHCALLY EQUIPPED BEINGS.** **That's what our common goal is!**

Every System Needs a Long-Lasting Structure!

5. Standard Education is Just an Initial Seed of Knowledge

On top of new knowledge, however, as early as possible, you also need to consciously expand their general outlook and upgrade their intellectual self-development and self-actualization, mastering **SELF-EDUCATION** in **TEN VISTAS of INTELLIGENCE** *(Part Three below)* as the essential ones for personality formation, especially at **START OF ACADEMIC EDUCATION** that you need to keep enriching the entire life.

Being the Best is a Tough Test!

The AI time demands we supplement the standard education with the ever-changing, increasingly **objective vision of life**, backed up by a clear-cut idea how to **SELF-INSTALL** in it professionally and personally, and how to realize the unique potential that often remains an unrealized dream in a randomly lived life.

The time for intelligence to prevail has finally come! Our young people need the **BLUEPRINT OF ACTION** to be mastered for the rest of their lives. It must be done by adjusting to the changing times because the mind cannot change the brain, ***but we can!*** Knowledge becomes true power only when ***you continue to self-educate exponentially, inducting yourself with:***

I CAN….! I WANT TO…! And I WILL…!

The book forms the blueprint for action as the **MANUAL** for **SELF-MONITORING** and **SELF-MENTORING** in five dimensions. Holistically formed intelligence will create in you a ***new holistic memory bank*** **,** cleansing off the redundant, outdated, and messy information that you might have accumulated for years of impulsive, automatic, and unconscious learning.

New intelligence, fortified by *Artificial Intelligence* will transform us into future **TRANSHUMANS** that will manage to put the **MIND** and **HEART** in sync to finally obtain the connection with the *Universal Mind* in a godly, consciously intelligent, and holistic fashion. You will have every right to declare:

In My Life Quest, I'm the Best!

There wasn't, there isn't, there won't ever be Anyone like Me!

Maintain physical, emotional, mental, spiritual, and universal hygiene.

You are the Boss of Your Own Self-Worth!

Resist the Gravity of the Common Thought.

Learn to Be Equal to God!

"It is Not Enough to Be the Best. Be the Only!" *(Steve Jobs)*

Part Two

Information Systematization

Re-Educate the Mind to Become Self-Refined!

Your Mind's Potential is Exponential!

"The Whole Science is the Mental Refinement of Day-to-Day Thinking in all its forms." *(Albert Einstein)*

(Best Pictures / Internet Collection)

"The gift of mental power comes from God, and if we concentrate our mind, we become in tune with this great power." *(Nikola Tesla)*

You Are HOW You Think!

1. Mind Means Intelligence + Self-Consciousness!

Rationalization of life in five dimensions is the first step in the paradigm of Self-Resurrection. The human body is a mental energy conduit. We need to learn to channel our mental energy in ***sync*** with the universal energy of creation that is **MENTALLY DIGITAL** according to the newest discoveries in physics. We are governed by ***Universal Laws*** that need to be studied and observed without negligence to be better coordinated with them.

Our Goal is to stop being Universally Negligent and become Universally Intelligent!

"Without better intelligence, without acute awareness, without knowing what we really are, we cannot refine even one cell in our bodies." *(Don Miguel Ruiz)* The quality of our life is the quality of our cells, our biological computers that **monitor and mentor our inner life.**

Dr. Bruce Lipton, an advocate of Informational *Medicine,* suggests we re-program ourselves at the cellular level ***to enhance the quality of our self-consciousness, which is our reference point***! As mental creatures, we must be always aware of our merging with the "***Universal Energy Field***" *(Dr. John Hagelin)* because we have the divine purpose **to become One with Universal Intelligence.** The body and the mind *(form + content)* are the vessels of our immortal souls.

Accumulating ***Holistic Conceptual Intelligence,*** we_should sort out carefully the information that we get from various sources and see what suits the body and mind best. We cannot go with the flow of mass media information blindly. Rationalizing life means ***being autonomous in thinking*** and becoming more self and life aware and sincerely responsible for both.

SELF-MENTORING and **SELF-MONITORING** are an integral part of Self-Resurrection! ***"Know thyself!"*** maximum must govern our progress at each level of the holistic paradigm and in every dimension. In other words, we should,***" practice mind over brain attitude"*** *(Dr. Rudolph Tanzim*) It means having the mind, devoid of inner conflicts and obsessive, impulsive reactions to the past or the present, the mind that works in sync with the heart, constantly dictating to you,

If we develop the ability to monitor the permanent **MIND+ HEART** connection in making any decision and solving any problem in both our private and social life, we will learn to live responsibly. The disconnection between the heart and the mind means death of relationships, marriages, friendships, job objectives, and any life expectations. Going after one or the other does not raise our self-consciousness or establish the soul-symmetry of the form+ content of life.

(Body+ Spirit+ Mind) + (Self-Consciousness + Universal Consciousness= Soul-Symmetry!

Prove Your Exceptionality without Personal Vanity!

Intellectualize the Heart and Emotionalize the Mind.

Be One of a Kind!

2. Living Intelligence for Us and Humanoids' Mass

Our evolutionary development on the mental level calls for us to become more self-aware, creative, self-confident, self-reliant, and self-programmed. ***It is not an automatic process; it's a reflective one*** because the mind analyses life constantly and channels us along the evolutionary steps of raising our consciousness if it is consciously directed that way.

The digitized mind must be bold to fit in the evolutionary mold!

We are overwhelmed with information, and therefore, our attitude to reading and self-enriching should change radically. I think that beyond the absolute necessity to get familiar with the basics of *philosophy, logical thinking, classic literature, music, art, jazz., etc.,* our **SKILLS OF COGNITION** should be electronically enhanced.

The ability to present, process, and store information needs to be as close to that of humanoids as possible. The purpose of our aware attention must be REFINING OUR HUMANNESS, not just humanizing the machine.

It is the time of our information filling and self-reformation fulfilling!

This main skill cannot be developed by writing a diary, reading a sentimental novel, gaining life experience from superficial , experiential stories about sex, ("*sex sells*") crime, and the negatively charged records of a psychiatrist's clients from self-help books. **The time for such literature is gone!** The genres of the publishing industry need to be reconsidered for their time validity.

Digital intelligence alters the way we think, write, and speak, and "***at the cellular level, technology is revealing how a person's neurons update themselves to generate a living library of experience***."(*Scientific American Mind, Nov./Dec.2016*) Obviously, we need the new skills - the **INFORMATION PRESENTATION** and **INFORMATION PROCESSING SKILLS**, *(See the Mission Statement above)* , enabling us to quickly process the information for its validity, sift the redundant noise in it, and store only those chunks of information that have associative connection with the bits of the previously selected knowledge. We should not clog our brains with informational negligence!

Aware, attention should never be in retention!

We need Purposeful Intelligence now!

The Center for Assessment Technology and Continuous Health (CATCH) is developing tools that can continuously monitor people to make electronic intelligence better serve our needs. Artificial Intelligence is dead and programmed, but it is our present and the future that is a human + machine mutual. ***Self-Emancipation is still your elation!***

I Am Good! I Exude My Self-Management Mood!

We are God-Created, Not Machine-Mind Mandated!

3. Strict Mental-Emotional Hygiene Must Be on the Scene!

Giving more tribute to our fantastic computer scientists, engineers, and designers, let us remember that the brain "***invents entirely new coding schemes for novel situations.***" *(Michael Freedman)* The ability of such intelligence is called," ***the improvisational dexterity of the brain."***

Holistically power up your inner digitized rehab!

An American psychologist ***Richard J. Haier, k***nown for his work on neural basis of human intelligence writes, *"We need to set up an entirely separate set of neural signals, create implantable brain chips that can restore or enhance memory thanks to the embedded electrodes in slices of hippocampus*

"It is both liberating and constraining in its essence."

This is an incredibly challenging goal because this process requires strict mental-emotional hygiene to be observed by us with the help of meditation and auto-suggestive **SELF-AWARENESS** .Also, we need to learn **INNER BRAIN SCANNING** that requires at least the dilettante knowledge of the brain structure and its functions that I keep accentuating all along.

If a robot designer knows what sections of the machine brain he is working on, he also needs to know what sections of his own brain are responsible for his actions and how to make them both work better in sync.

These techniques should be studied in schools **in partnership with thinking machines**. It presupposes the necessity to develop the habit of sorting out and deleting information that is polluting the brain to organize it into the **CONCEPTUAL CHUNKS of SORTED OUT KNOWLEDGE** at the conscious effort of the learners. The brain needs to be able to compartmentalize the incoming information holistically and connectedly at the command of the conscious mind as opposed to a computer that does it automatically.

The indispensable AWARE ATTENTION SKILLS must be developed in learners.

Mental discipline is the skill that needs to be installed continuously and consciously, still! Let's give much credit to the producers of ***space trip movies and TV shows*** that promote extra-terrestrial thinking, making us marvel at the scope of their intelligence that expands the unused capacities of a human brain. *The Oscar award movie* "**Gravity**" is a great illustration of what we might be like in the future. Our great actors, *Sandra Bullock and George Clooney,* materialize our future thinking capabilities in an overwhelming way!

So, it is paramount for us now **to polish both sides of the coin,** *the Moon, and the sixpence*" sides ("*The Moon and Sixpence" by Somerset Maugham*) with the priority placed on the values of ***the Moon side*** of the human intellectual development not ***the sixpence side*** of our digitized development for ***lucrative, life-perfecting, but soul but soul-destructive needs.***

The Quality of Our Thinking Becomes the Priority!

4. Human +AI = Holistic Intelligence Network

Thus, Digital Intelligence is ruling our intelligence on the ladder of our AI backed up Self-Installation. Meanwhile, we are governed today by mechanical intelligence and boundless irrationality that information turmoil create in us. ***Our transhuman*** *(Ray Kurzweil)* ***transformation*** or our merging with the machine is happening now **in a very chaotic way**. This process needs the **KNOW-HOW** and the **PLAN OF ACTION** that are main educational objectives now and my attempt to provide my Psycholinguistic methodology of treating this problem holistically make up the essence of this book presenting the mental level in a five-dimensional paradigm of our holistic self-growth.

Brain Science + **Psychology** + **Languag**e = **SELF-CONSCIOUSNESS DEVELOPMENT!**

Our fantastic designers have found a way of connecting human + Humanoids neurological systems algorithms. I am sure special applications *(brain chips in exceptional cases)* will be created to make brain devices removable, without a life-defining needs to implant brain chips . Thu AI will contribute immensely to enhancing our imagination, expanding the outlook, and building up a new **INTELLIGENCE NETWORK** - the future **BRAIN-to-BRAIN NET** ***(*** *Dr. Kaku* **).** It will serve for our communication with each other and with humanoids at a thought-movement, used as a mouse. How can one remain indifferent to this well of knowledge and to the scope of intelligent minds that we can connect to from all over the world? It is, indeed, very enriching, and life-span lengthening.

I wish I could live then in the answerable When?

The Internet informs us, for instance, that a group of scientists, working on the phenomenon of our DNA, has discovered lately that it is able to ***produce magnetic black holes*** as miniature bridges to transmit information.

According to the Russian scientists that are trying to probe further ***a genius supposition of Einstein-Rosin***, "**these bridges might be connecting different areas of the Universe,**" letting the information be transmitted beyond the time–space continuum. They are trying to discover how to consciously monitor these communication lines, ***using our DNA as a receptor***, able to transmit and receive the information from the **UNIVERSAL WEB WIDE NET**.

What an incentive to live and to create beyond ANY cultural and political differences! Wonders of technology are being developed all over the world at an unprecedented rate changing the intellectual picture of the world in every field of knowledge.

They globalize the economy and **remove the barriers between the science fields** and interpersonal communication. Our social consciousness develops with new scientific ideas steering the world of thought.

Life is a Dynamic Interplay of Consciousness, not just Digitally Enhanced Intelligence!

5. Holistic Enlightenment is a New Frame of Thinking!

In sum, this most advanced attitude to AIs needs ***to be human-monitored, both consciously and consistently.*** We should be aware that these ***new modes of patterned thinking and behaving*** are occurring within the framework of a person's intellect, without limiting it, and indoctrinating it in the programmed direction of the same " **COLLECTIVE UNCONSCIOUS**"(*Carl Yung*)that is digitally enhanced now We are experiencing **DIGITA**L **ENLIGHTENMENT** now with the new breakthroughs in physics, biology, neuroscience, medicine, genetics, and electronics. New knowledge reveals the **scope of our ignorance** about the amazing structure of our bodies that, as it appears, is connected to *Universal Intelligence*, enveloping us everywhere, digitally. *". (R. W. Wetherill)* In other words, we must complete five stages on the path of complete **SELF-ACTUALIZATION**:

Self-Awareness + **Soul-Refining** + **Self-Installation** + **Self-Realization** + **Self-Salvation.**

The route of **holistic self-growth** is consequentially going from *Self- Knowledge-+ Emotional Self-Control + Knowledge - Internalizing + Self-Externalizing + Self-Enlightenment* to the time-demanding technologically enhanced

SELF-ACCULTURATION

It must be mastered by us ***from top to bottom****, from the universal level to the physical one, following the ancient maxima* **"As it is Above, so, it is below!"/ As you think , so you are!**

1. Enlightenmen*t* at the **universal level** *means the spiritual growth of a human being in an unbreakable unity of the* ***form and content of life in its entirety****, obtaining* **SELF-SYMMETRY.**

2 Enlightening in the **spiritual realm** *is the process of* ***spiritual upgrading of self-consciousness,*** *limiting or eliminating completely our still boundless religious ignorance. We need to focus on co-****creating ourselves*** *in the limited time and space that each of us is granted from the Above to complete our mission on Earth.*

3. Enlightenment in the ***mental dimension*** *of our holistic development means proper digestion of the incoming flood of information and its assimilation that require our conscious interference into* ***the chaotic process of automatized, unconscious life. At present, our being determines our self-consciousness. But it should be the other way around:***

Our Self-Consciousness Should Determine Our Being!

Developing self-consciousness and eventually connecting it with Universal Consciousness is our **emotional** and **physical** goals if we want to become ***Beyond the Terrestrial*** people with the mission of "***colonizing other planet***" *(Elon Musk)* and enriching them with our ability to love.

"We need to Evolve Ourselves for this Mission. *"(Elon Musk)*

(End The Information Systematization Part of the Book)

Spiritualized Intelligence + Self-Consciousness

= ***Living Intelligence** or **Self-Developmental Paradigm***

Form — **Content**

Reformed Habits + **Technologically Enhanced Skills**

Self + Life -Awareness — **Cause-Effect Reasoning**

Right Thinking — **Right Speaking + Feeling+ Acting**

Spiritualized Intelligence — **Self-Consciousness**

= ***A Whole, Intellectually Spiritualized Fractal of Self!***

. "The Unexamined Life is Not Worth Living!"

(Socrates)

Part Three

(Mind-Refining Part of the Book)

Intellectual Odyssey of a Seeking Soul!

(Five Dimensions of Intellectual Growth)

Use Your Intellectually Spiritualized Self-Gravity!

Holistic System of Self-Resurrection.

"Fertilize Your Mind with Thinking!"

(Albert Einstein)

"I Think, therefore, I AM!"

(Descartes)

Ten Essential Vistas of Intelligence

(Two types of intelligence at each level of Self-Resurrection)

I know who I am and Who I am Not!

I am a strong, calm, bold, and determined owner of my solid will!

I Can…! I Want to…! And **I Will…!**

"Life is a Flow of Intelligence. I Take All My Ideas from it." *(Nikola Tesla)*

1. Soul-Symmetry Formation

Self-Awareness + Soul-Refining + Self-Installation + Self-Realization + Self-Salvation = Intellectually Spiritualized Self-Actualization!

Enlightenment in five realms of life makes it increasingly apparent nowadays that we need more "***scientific literacy***" and holistically applied new knowledge for **DIGITAL ENLIGHTENMENT** to be successful.

"**The *world of cells needs to be studied because it will help us understand the fractal nature of cells that have their own individual life*"** *(Arthur C. Clarke)*

We need to be much more informed about life on the cellular level that needs to be re-programmed, following the mind-blowing discovery of the fractal nature of life" *(Dr. Mandelbrot).*

Following *Dr. Mandelbrot's* breath-taking discovery, I see the process of our Self-Acculturation now in the formation of our own spiritual fractals of Self-Symmetry in us as:

Universal Level — **Oneness** (*Inner Oneness + the unity with all life)*

Spiritual level — **Grace** *(Conscious faith, nobleness)*

Mental level — **Mind** ***(S****piritualized intelligence)*

Emotional level — **Spirit** (*Blissfulness, willpower)*

Physical level — **Body** (*Energy, wellness)*

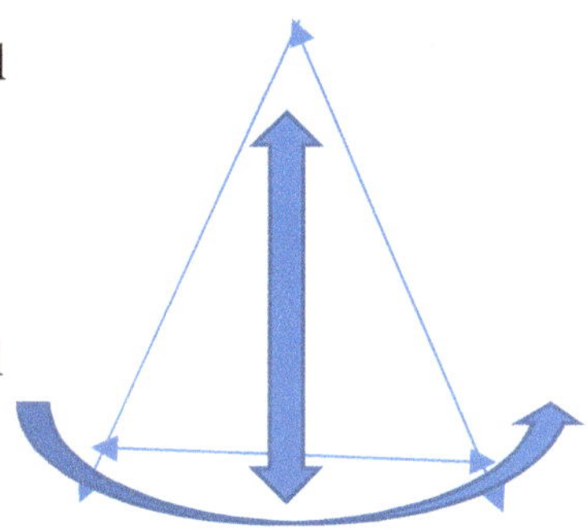

(Body+ Spirit+ Mind) + (Self-Consciousness+ Universal Consciousness) =

An Enlightened, fractally Whole You!

There is a new branch of medicine, called "***Informational Medicine***," led by ***Dr. Bruce Lipton*** that I have quoted above. *Dr. Lipton* claims that our cells, not just DNA, determine our lives, and if we learn how to program and re-program them as our minicomputers, we will be able to improve our health, change our character, and become true heads of our bodies and lives

It is also vital that we acquire more knowledge of the brain functions and more awareness how to operate it consciously, so we could work out our own knowledge-enhanced, individually processed grid of experiential perception of life - our own self-monitoring techniques. Quantum Computation is a new stage of our technological revolution, and the advancement in Quantum Computing + Generative AI are breathtaking in every field of knowledge with human evolutionary transformation is on top of these developments. .

Our Omega Point is Not Retirement; It's Enlightenment!

2. Accumulate Holistic Conceptual Intelligence

"Life is Not a Circle. It is a Spiral!"*(Dr. Fred Bell 0*

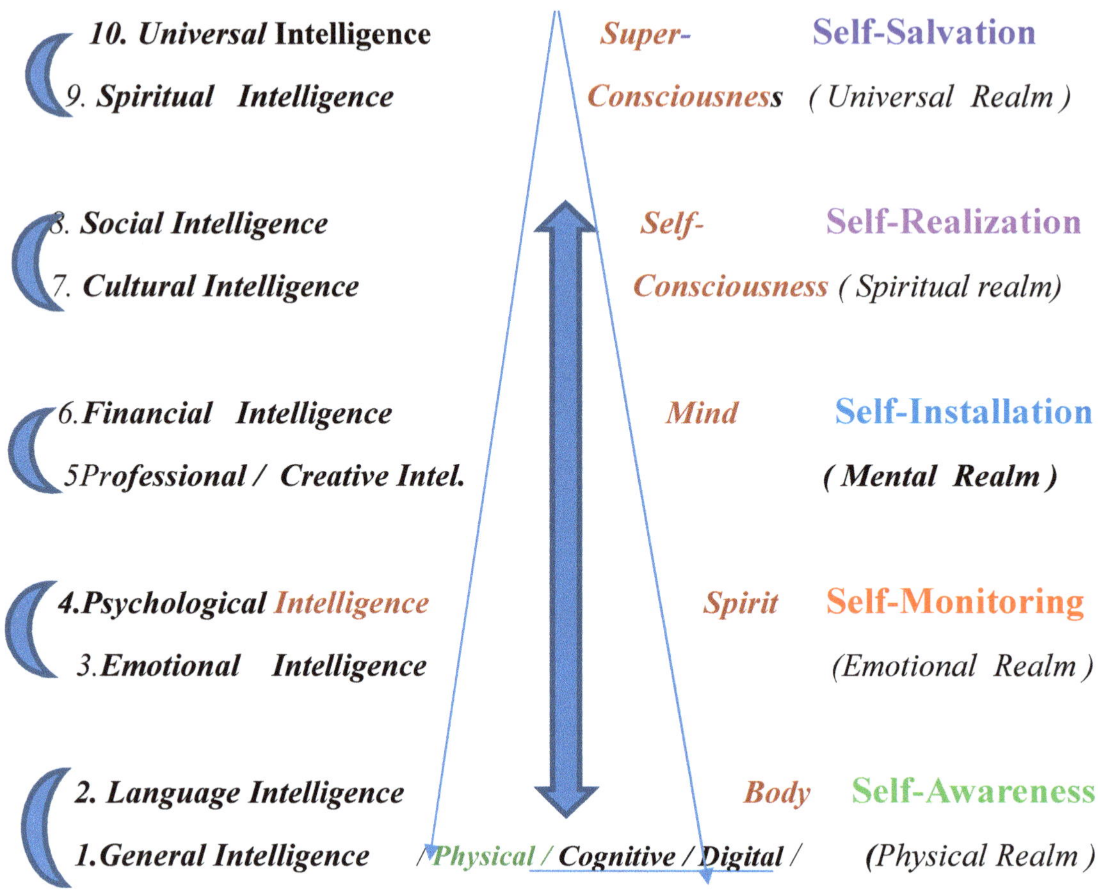

***(Body +Spirit + Mind +Self-Consciousness +Super-Consciousness)=*A Whole You!**

The conceptual framework of the **Holistic System of Self-Resurrection**, ***presented as the Mission Statement of the book*** reminds of the structure of the Russian dolls, ***Matryoshkas,*** when one level incorporates the next one, forming one simple holistic system of self-installation in life, prompted by the present-day digital upheaval and the necessity to adjust a personal pace to it in the ***physical, emotional, mental, spiritual, and universal realms of life.***

Thus, we are ***forming*** "**Singularity**" and heading toward **BRAIN-TO-BRAIN** interaction between us and the machine. The chaotic state of our ***mental education must be holistically systematized and strategized toward a learner's full Self-Realization.*** Our memory banks should be loaded with time-relevant and scientifically verified information.

The schematic structure above presents the* INTELLIGENCE REFINERY *in ten vistas of intelligence outlined consequentially below in ten stages.

Inflate Your Personal Pride with the Intellectual Self-Might!

3. The Spiral of Self-Resurrection Must Be in Your Intellectually Spiritualized Self-Reflection!

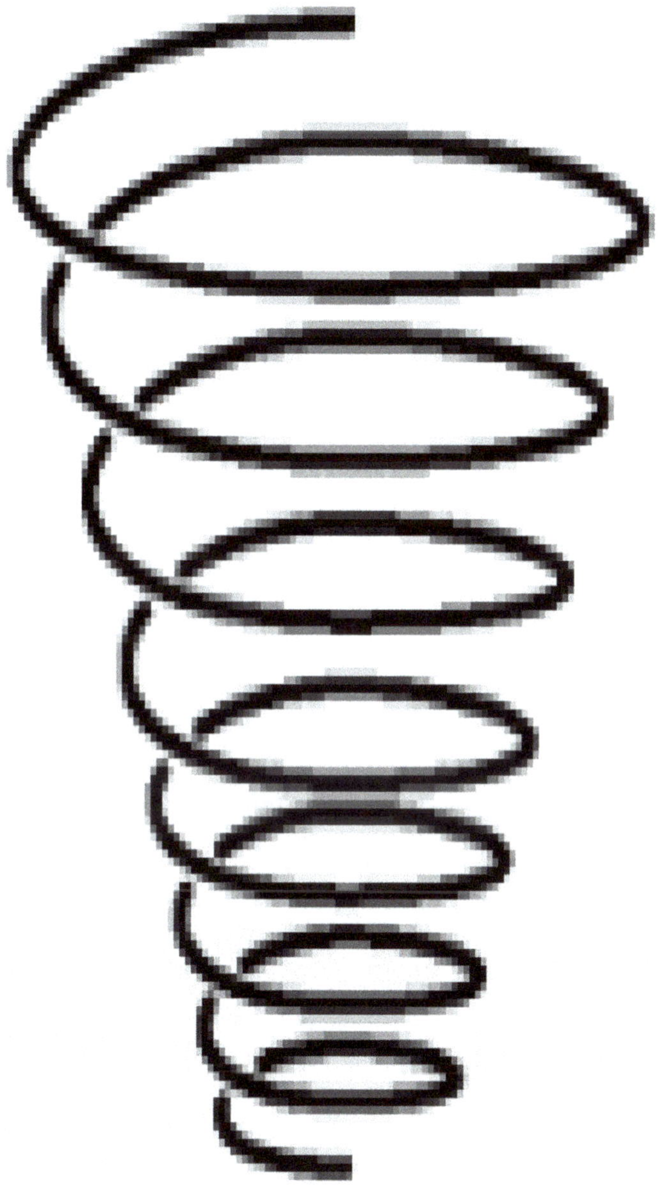

"Life is not a circle. Life is a spiral!"

(Dr. Fred Bell / "Death of Ignorance")

The world is moving to ***a new order,*** and we must act steadfast, reenforcing the foundations of our **HUMAN + AI CODE of THINKING** that is changing in a spiral way, uniting ***science + religion +AI's transformation*** that is supposed **to break our cicling stagnation** and push our intelligence development in the spiral mode toward **SUPER-INTELLIGENCE**.

Our Developmental Function is to Become God in Action!

4. Choreograph Your Self-Creation with Elation!

Every year, before a New Year starts, we make our provisions based on what is essential for us in ***the Master Plan*** of our lives. However, very often, our mission statements do not guard our inner transformation because ***they are not properly structured strategically and tactically.*** They are not integrated into one, ***five-dimensional plan of action*** that you need to install into your ***Visual Cortex*** as the spiritually internalized direction, serving as the GPS for your self-growth.

In our life's ado, we are what we perceive, feel, think, say, and do!

To fill up your life with constructive action in five life-dimensions, conduct a quick **SELF-SCANNING** before sleep in the *physical, emotional, mental, spiritual, and universal dimensions.* Synthesize your actions during the day and give yourself a general grade for the day.

Then analyze your accomplishments in each realm of life separately, also assessing your progress or regress in grades. Finish your self-scanning by **synthesizing** the steps that you need to take the next day. Let the paradigm **Synthesis-Analysis- Synthesis** choreograph your life by integrating ***the form and content*** of your life to attain soul-symmetry. *("Soul-Symmetry"/You Tube video)*

(Body+ Spirit+ Mind) + (Self-Consciousness + *Universal Consciousness*)

Please, note that **the stages of Self-Resurrection** - **Self-Awareness, Self-Monitoring, Self-Installation, Self-Realization,** and **Self-Salvation** unite you into **ONE -YOUR LIFE!**

Visualize the structure of ***five Russian dolls - Matryoshkas*** that go one inside the other with the Mother Doll on top. These stages also unite you into ***One - a unique human being.***

"***To attain knowledge, add things every day, to attain wisdom, remove things every day.***"

(Lao Tzu)

You need to design and choreograph every Pa in our "life ballet!"

Our ***Divine Books*** in any religion teach us **Self-Salvation**, but they remain a side dish in our minds when it comes to action because we do not have a clear-cut picture or the **PLAN OF ACTION** in the mind. ***The Law of the Right Human Behavior*** (*"Right is Might!" by Richard Wetherill*) lays out **the what, the how, and the why** we need to work on self-perfection with self-reflection.

Humanoid's' neuronal system should also be designed to analyze his /her "human" qualities... The gift of consciousness granted to us demands we do our own **INDIVIDUAL MOLDING** that can be backed up by a specially trained robot friend. We live under the laws of God that "***the Celestial Court instills in us***" *(Robert Stone*) and that should not be pushed aside for a better time. **"The right time never comes!"** *(Osho*) The exponential growth of the AI demands that we should better make it Now. ***Then, the wonders of the Future will make us and AIs more mutual!***

No Time to Spare. Beware!

Resist the Gravity of the Common Thought.

<u>Magnetize Yourself with What You Are Not!</u>

(Pictures of the brain are by Yolanta Lensky / My daughter)

To Life-Sustain, Help Your Mind Monitor the Brain!

Stage One – Self-Awareness

*Mini-level –**Physical Dimension** - Digital Psychology for Self-Ecology*

General

Intelligence

*(**Physical, Cognitive, Digital** divisions)*

Accumulate the Spirit of the Five-Leveled Self-Installation with a Digitized Self-Awareness and Life-Elation!

The Intellectual Odyssey of a Seeking Soul

Vistas of Intelligence to install next:

Let AI Enforce Technological OutBurst!

10. Universal Intelligence
9. Spiritual Intelligence

Self-Salvation
(Universal Realm)

8. Social Intelligence
7. Cultural Intelligence

Self-Realization
(Spiritual realm)

6.Financial Intelligence
5. Professional /Creative/Intelligence

Self-Installation
(Mental Realm)

4 Psychological Intelligence
3.Emotional Intelligence

Soul-Refining
(Emotional Realm)

2. Language Intelligence
1.General Intelligence

Self-Awareness
(Physical Realm)

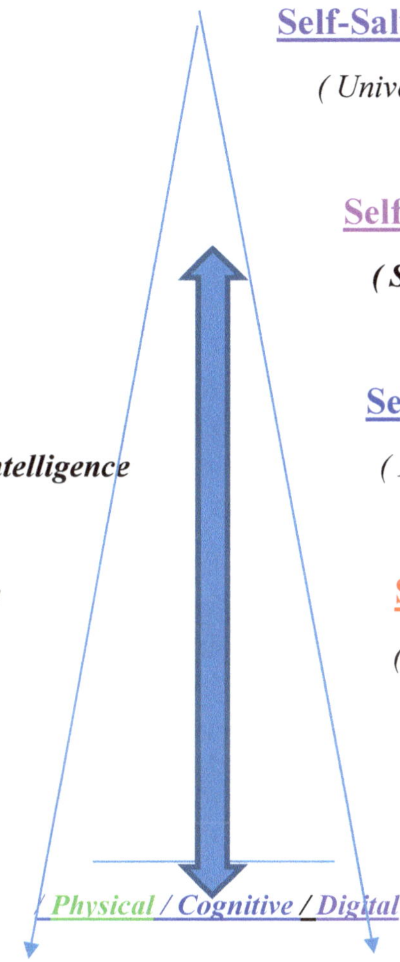

Cleanse Your Mind with a Digitized Brain Rewind!

To Make Your Life Digitally Wise,

Generalize - Internalize - Personalize - Strategize –Actualize!

Don't Be Generally Negligent. Be Generally Intelligent!

1. Be More Aware. There is No Time to Spare!

To begin with, we need to get rid of the chaotic and outdated knowledge about life and living and change it into ***holistically systematized and well-organized banks of knowledge*** that will enable us **to control the humanized machines** that are getting self-aware now. Our future interaction with them necessitates a considerable enrichment of our intelligence in five dimensions - ***physical, emotional, mental, mental, spiritual, and universal*** with two vistas of intelligence in each of them These five dimensions are ***objectively structured,*** but the knowledge about each vista of intelligence, representing them *(See above)* are **subjectively accumulated** by each of us . Information that is relevant for you might be redundant for someone else. You need to constantly update the chosen information with ***scientific enlightenment*** that must make up the basic core of your holistic "***scientific literacy***".*(Dr. Neile de Grass Tyson)*that every educated person should be developing now.

Holistic Conceptual Intelligence must be accumulated without negligence!

We all need intellectual refilling and spiritualized stewarding! ***S***o, general intelligence incorporates all ten levels of intelligence into **ONE SYSEM of CONCEPTUAL INTELLIGENCE** that we call **IRUDITION.** Conceptual knowledge is always organized or compartmentalized in accordance with our professional or personal needs, general education, and outlook.

General Intelligence allows a ***Self-Acculturation-minded*** human being to set new, well-organized memory banks that store the time-relevant and subjectively needed chunks of information, enabling a person to accumulate **HOLISTIC INTELLIGENCE** in different fields of knowledge. Humanoids have such depositories, and they provide the information that we need with mind-boggling speed. We do not need to feel inferior to them because of that. **They are our employees of the mind. Let them do information rebind.**

The accumulation of **HOLISTIC CONCEPTUAL INTELLIGENCE** in five life dimensions will help us form **quality intelligence,** closer to that of humanoids. Human intelligence is much more intricate and ingenious in our ability **TO MOLD** intelligence in the holistic way to apply it better and more reasonably to any situation by the paradigm - ***Synthesis - Analysis – Synthesis.***

Generalize - Analyze - Internalize - Strategize –Actualize!

Artificial brains (*left and right hemispheres*) of humanoids are designed to be in perfect sync, but their neural net, Amygdala glands that are uploaded with a sophisticated emotional array, will never be equal to ours. Humanoids might display a set of different emotions, but ***they are not in an integral unity like ours***, and therefore, their machine ability to love and be loved in return will never surpass ours. Naturally, our primary position in line requires intelligence, enabling us to conduct **CONCEPTUAL ANALYSIS** OF any physical, emotional, mental, spiritual, and universal situations. **So, become a Jack of All Trades and master of All!**

The Time for Professional One-Sidedness or even Two or Three Majors in a Diploma is Gone!

2. Digital Psychokinesis in Action!

Auto-Suggestive Psychology for Self-Ecology *(both inspirational, digital, and quantum)* that I use in all my books is extremely helpful in learning anything because it is focused on **STRATEGIZING** *physical, emotional, mental, spiritual, and universal realms of life in us holistically***.** It is, in fact, ***the process of psycho-kinesis*** - **raising of one's own spirit at will!**

The aim of such psycho-persuasive culture is the development of the ability to make one's own thoughts and emotions "***I-directed***. *"(Dr. Noam Chomsky)*. Conscious , self-directed, self-boosting inspirational programming enables a person to actively be his own textbook and his own mentor, his own psychologist, and a judge. In a self-training program, you transform your own consciousness with **AWARE ATTENTION**. *(See the book "Self-Taming / Spiritual realm)*

Auto-Induction: **I am the manager of my own framing and re-framing!**

We are all familiar with the positive effect of affirmations and quotes, but auto-suggestions work even better because they are ***directly targeting you and only you*** in a simple, rhyming and most persuasive way. *You manage to re-program your habitually negative, depressive thoughts and emotions to the positive and constructively charged ones* in simple, rhyming words. Thus, you inspire or fuel yourself on this path, raising your **immunity to the negativity of life,** both internally and externally

*"**A rhyming word is going better inward!"** (Edgar Cayce)*

I witness amazing changes in my students, once they start ***auto-suggestively altering their minds knowingly and holistically.*** Their personalities get revived like a drying plant that you water to save from dying. Research shows that individuals differ in the quality of their vibrations which constitute their individual personalities. *Mr. Birched* writes about it, "***We have a top–down mechanism for managing our bodily mechanism. In modern humans, we are learning to operate outside our consciousness, analyzing our existence from inside out and from outside in"*** ("*New Scientist," Aug. 2913)*

Self-training through self-induction is the best way for self-production!

You are the one who channels your mind and operates your emotions! However, the faculty of consciousness which many people lack is successfully developed only in those of us who try hard to accomplish a higher level of consciousness in sync with **AWARE ATTENTION** paid to learning.

In his wonderful book *"**The Ravenous Brain**," Daniel Boor* writes, ***"Consciousness is effectively an idea factory – a key component of which is the discovery of deep structures within the content of awareness of our own imperfections*****.***"* Life is full of ideas if we can perceive them.

Love of Life, Will-Power, and Discipline are on the Digitized Self-Actualization Scene!

3."Dissipated Consciousness is a Wasted Life."

The famous words of *Carl Yung* above characterize the state of our self-consciousness and that of the humanized machine – beings now. ***We are not self-conscious. We are not*** **CONSCIENCE - CONCERNED.**

WE ARE IN A SELF-CORRODING FORT!

We are degrading ethically , morally, and personally, taking the wonders of technology for granted. The wonders of Artificial *Intelligence* are far from perfect, and therefore, they constantly declare their intention to destroy humanity. So, the goal of the present-day education should be the creation of the ***Holistic Conceptual Intelligence*** in us through **systematization and simplification of knowledge** that needs to be time relevant to secure the needed "***scientific literacy"*** *(Neil de Grass Tyson), on the one hand* , and inspire learners to develop personality-wise and accomplish full self-realization in life, *on the other*. **We need self-perfection, AI conducted sessions.**

Don't teach just the subject. Instruct the whole person!" *(Leo Vygotsky)*

Let me remind you that this book is backing up **the third, mental level** of ***the Holistic Paradigm of Self-Resurrection***, featured in the *physical, emotional, mental, spiritual, and universal* levels of self-growth. It was originally presented in the *Excellence Book Award Winner, "Living Intelligence or the Art of Becoming."/ Inspirational psychology,* and this book is its up-dated version. The exponential growth of Artificial *Intelligence* has necessitated establishing control over the wonderous creations of humanized machines and our training of them in the holistic paradigm ***Synthesis-Analysis - Synthesis*** and in five dimensions of life together . It is the way to substitute our common ethical ignorance with true **GRACE + NOBLENESS + LOVE** authenticity. ***Together, we will be creating intellectually spiritualized fractals of "Singularity" effect on us, forming a*** **HOLISTIC BEING** ***that is becoming*** *aliens -like transhuman being.*

(Body+ Spirit+ Mind) + (Self-Consciousness + Universal Consciousness)

Visually backed up holistic thinking will help us track *Artificial Intelligence* scientific and technological progress and inspire us to alter our irrationality, impulsiveness, and fun chasing, creating gradually the domain for a new **WHOLE BRAIN**. Psychologically aware **SELF-MONITORING** and **SELF-MENTORING** are the ways for our physical transformation, enhanced with **AI's psychological supervision**. A new memory bank, ***formed holistically in five dimensions of life*** and enriched with ten essential vistas of intelligence *(See above)* will help us construct **HOLISTIC CONCEPTUAL INTELLIGENCE** that must become our digital tank in which we will be victorious on the battlefield for superiority over humanized machines . If we provide them with new ethically enriched data with the help of ***psychologically based applications*** to modify our ignorance and negligence of this plane, we will improve our common ethical negligence. No nagging, no guilt-trips, no yelling, or reprimanding.

Generally Shared Knowledge will become Our Common Ethical Storage!

I Am Unique in Every Stance. I Was Born, but Only Once!

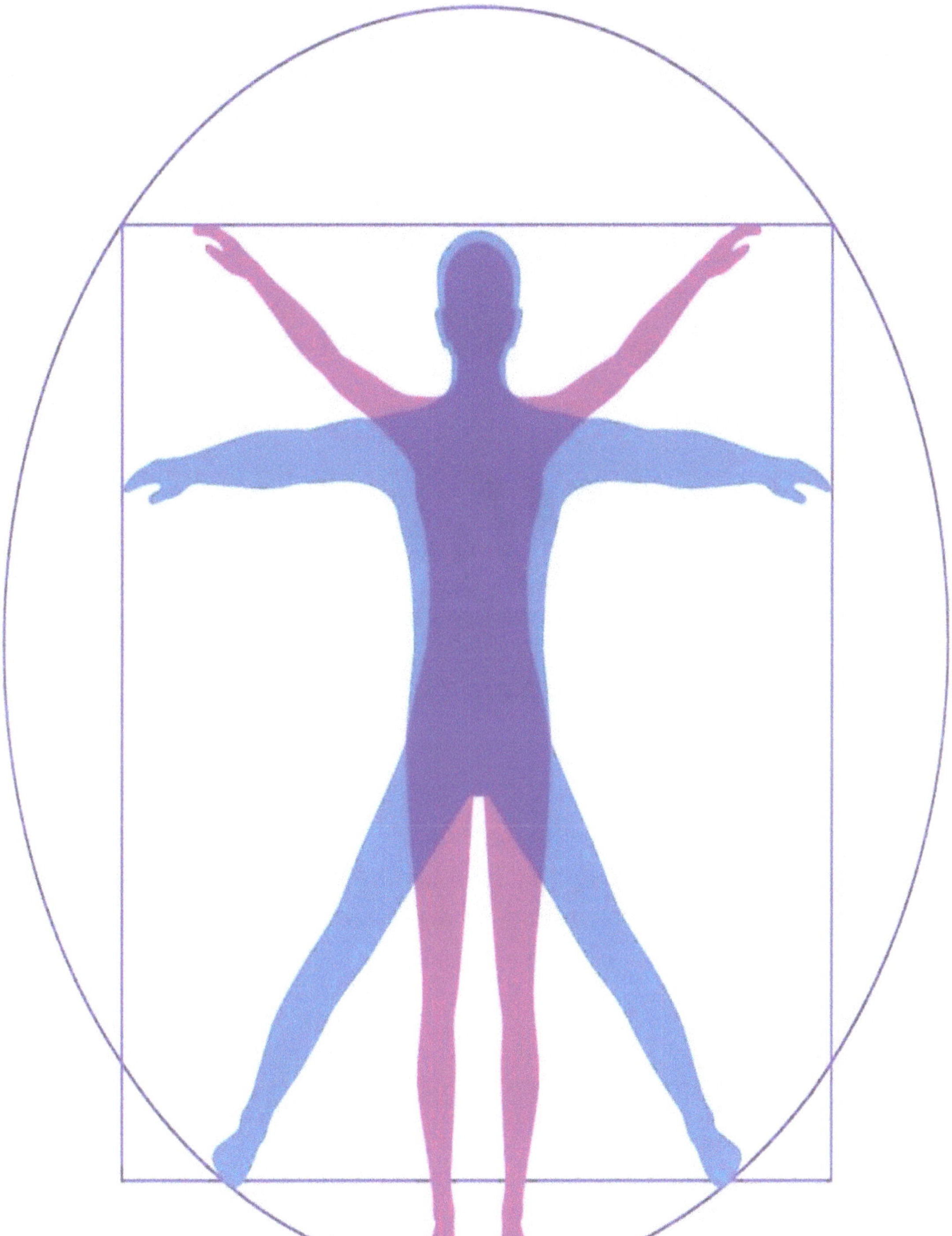

There wasn't, there isn't, and there won't Ever Be

<u>Anyone Like Me!</u>

Stage One, Division 1

General Intelligence - Physical Dimension- Self-Awareness

Physical

Intelligence

(Primary Self-Genesis)

(For more on the physical dimension of Self-Creation, see the first book on the Holistic System of Self-Resurrection -- ***"I am Free to Be the Best of Me!"***

Inflate Your Personal Pride with New Intellectual Might!

The Intellectual Odyssey of a Seeking Soul

Vistas of Intelligence to install next:

Let AI Enforce Technological OutBurst!

10. Universal Intelligence	**Self-Salvation**
9. Spiritual Intelligence	*(Universal Realm)*
8. Social Intelligence	**Self-Realization**
*7. **Cultural Intelligence***	***(Spiritual realm)***
6.Financial Intelligence	**Self-Installation**
*5. **Professional /Creative/Intelligence***	*(Mental Realm)*
4.Psychological Intelligence	**Soul-Refining**
*3.**Emotional Intelligence***	*(Emotional Realm)*
2. Language Intelligence	**Self-Awareness**
1.General Intelligence	*(Physical Realm)*

*/ Physical / Cognitive / **Digital***

Fertilize Your Mind with a Digitized Brain Rewind!

To Make Your Life Digitally Wise,

Generalize - Internalize - Personalize - Strategize –Actualize!

Don't Be Physically Negligent. Be Physically Intelligent!

1. Self-Emancipation is Real Salvation!

To monitor self-growth and self-modification, we need a new level of self-awareness that, in turn, demands a much higher level of intelligence. The beautiful words by Osho "**There are two ways in life to be and to know**" determine the conceptual structure of the physical intelligence part of the book that presents the holistic paradigm of self-creation at the mental level. Our primary **SELF-GENESIS** starts with figuring out **HOW TO BE** and **HOW TO KNOW** at the time of our digitized transformation *in the physical, emotional, mental spiritual, and universal realms of life*

Starting with the lowest level of the ten vistas of intelligence -*General Intelligence, **we need to . integrate physical, cognitive, and digital** levels.* Together, they build up a solid foundation for other vistas, or compartments of intelligence, that we need to install in our minds at the time of the exponentially changing technological reality.

You are shaped by the intellectual power that you have accumulated, selected, and applied. ***Be unique, not bleak! Discover your exceptionality without any vanity!***

"Low self-awareness is low-self-identity!" *(David Icke)*

"KNOW THYSELF!" We know this provision of God, but we are not fully aware of it. This situation is being generated by the technological revolution that is the greatest challenge for us , on the one hand, and a serious responsibility on the other. Fortunately, inner discontent and the search for balance in life push us forward, on the path of the **SPIRITUAL QUEST**,

Learning to be godly in a godless world of progressive de-humanization becomes increasingly difficult and time-consuming.

Physical, emotional, mental, spiritual, and universal growth is a material process, to begin with, analogous to the growth of a physical body, Fortunately, we have accomplished a lot on the path of bettering our bodies and helping them to be less strenuous in life.

BUT WE DID NOT BECOME ETHICALLY STABLE!

Our gains in the technologically emancipated physical intelligence are amazing now. We know the ways to better operate our life-diving body that is being duplicated in the thinking machines. We know more invaluable information about the body's functions, what energies it, and what for

We are trying to unravel the enigma of consciousness, but we cannot yet conquer our desire for material benefits first.

Our self-consciousness is not ripe yet, and we need to do the impossible to **SELF-CONQUER** ourselves to eventually become part of "***Universal Community of Light in which every nation is a separate ray if this light.*** " *(Nikola Tesla*)

"The Only Way to Discover the Limits of the Possible is to Go Beyond them into the Impossible!"*(Arthur C. Clark)*

2. A New You is a Better Aware You!

Developing at the physical level means a changed life and much better self-awareness. Awareness of the newest developments in brain science, digital biology, nutrition, and physical fitness of the body changes our limited view of life. I repeatedly mention that we all need "**SCIENTIFIC LITERACY** *"(Dr. Neil de Grass Tyson)* to wander the world with an open mind and a much more loving heart ***Revolutionizing our knowledge is the first stage of rationalization of our lives.***

"Let him who would move the world first move himself." (*Socrates)*

As we learn about the neural mechanism of intelligence, prospects for enhancing intelligence become inevitable. Your better self-awareness, disciplined thinking, and conscious action, backed up with added information that is sorted out for its validity help you strategize your life. Professional and personal **Self-Installation** needs constant professional self-control and intellectual **SELF-MONITORING.**

(Body+ Spirit+ Mind) + (Self-Consciousness + Universal Consciousness =

SELF-INSTALLATION!

I would like to recommend reading a wonderful book" The ***Start-Up of You"*** by *Reid Hoffman and Ben Concha,* the two great entrepreneurs from the *Silicon Valley*. The book is addressed to present-day young people with electronically charged minds who are consciously, intellectually and professionally transforming themselves into new transhuman beings.

"Be your own coach in life! Become a Superior You!"

Reid Hoffman, the founder of "***LinkedIn,***" is calling on you not to succumb to a machine but create more value in yourself and the world around you by **"*proactively investing in your intelligence, enhancing your entrepreneurial skills, and expanding your technical know-how."***

But such development requires a rigorous preservation of one's own individuality and unique way of thinking, without which there is no personal evolution. "***To run a successful start-up of you, you can and must think for yourself, create yourself, and invest in yourself!"*** *(Reid Hoffman)*

"Autonomy of thinking is a Must!"

The mind or intelligence is governing our life through its constant evolution that is downright impossible unless we uplift our mental and spiritual levels. The book "***Digital Binary and Human Refinery***" is just a modest attempt to holistically develop intelligence and up-life self-consciousness so that we could say in the outcome of life, "***No regrets***" or as *Mark Twain* wrote, ***"Let us endeavor to live so that when we come to die, even the undertaker will be sorry."***

I Know Who I Am, and Who I Am Not! That is My Transformational Fort!

3. Your Exceptionality is Your Personal Gravity!

Steve Jobs was the business genius of our generation and a real role–model for thousands of young, electronically-minded people because he had this exceptional ***self-developed intelligence and a keen sense of Self.***

Self-awareness propelled him on-ward, up-ward, and God-ward!

His professional integrity gave a go-ahead to the intelligence of many people, and he created a clear framework of opportunities and risky alternatives for the exploration of the artificial mind's possibilities that are endless. A great Russian psychologist *Leo Vygotsky* wrote,

*"**At the end of your life, all you are going to have been who you have become as a person. All is left for the people behind you. Leave a lot of yourself for them!**"*

When training robots, exceptional in their abilities that outpower ours, we need to pay attention to **personality development in a machine**, and its professional individuality, too. Since artificial clones are being produced now, they should also reflect the best qualities of a man /woman that they duplicate ***"The physical body is the temple for the mental one!"*** *(Edgar Cayce)*

So, "take charge of your internal pharmacy!" *(Paul Pearsall)*

All exceptionally talented people are known to be in constant search for ***the truth of life*** and their own special place in it. Obviously, there is no higher consciousness to be developed in us without obtaining higher intelligence that must be everyone's prior concern. Insightful self-quest with heart **+ mind** unity will uplift us to new levels of consciousness. Keep sculpturing the best of you holistically! ***Put heart and mind in sync. Feel, but think!***

The first strategic strength to acquire is the power of seeing, hearing, knowing, and reasoning at a much broader level professionally, culturally, and socially. But there is one enemy that we need to beware of. ***It is our constant battle with change and the desire for permanence***. But "**persistence beats resistance**!" *(Anthony Robins)* **Robots can teach us to stay on this path and beat our impulsive and irresponsible behaviors**. *However*, rationalizing our lives and perfecting ourselves is not enough for our "***intellectually spiritualized***" self-growth**.** *(Dr. Fred Bel)*

"*We also need to obtain connected consciousness with all!*"*(Schopenhauer)*

We have to learn how to be able to constantly ***harmonize our lives*** and get them ***in synch with the Universal Accords,*** changing the melody in our hearts to much more harmonious vibrations and becoming whole people that are able to experience ***the uniformity with all forms of life,*** harmonizing themselves from within and radiating light back to the world. We were promised to obtain this skill sometime in *the Golden Age* of humanity, ***but why not now, in our extremely limited time / space continuum?*** To be more life-fitting, keep repeating:

Drink the Juice of Life, But Do Not Drown in it!"

(Sadhguru)

4. "Man is a Free Agent!" *(Delia Lama)*

Live in God's Standards, not in People's Grandeurs!

Physical intelligence is the basis of our inner integrity, family values, and moral standpoints, instilled in us by our parents. It also questions the existence of a superpower over our life and the authenticity of our faith in it. With knowledge about the body, we acquire more awareness about the quality of our life, and we get physical maturity that helps us ***resist the mechanical reaction*** to life problems that operate on us when we are in a blind, unaware state of mind. As *Mr. Ouspensky writes,*

"There are only two forces struggling in the world- consciousness and mechanicalness."

Edgar Cayce in his very thought-provoking book "***No Soul Left Behind***" notes that every one of us has a talent for self-promotion –***"the energizing spark that lies within our soul***." We are all guided by the standards that are shaped by our intelligence and the cultural environment that shapes our values. *Edgar Cayce* also writes, "***Only constructive, well-formed intelligence may make for better physical and mental development"***

"Our cells in their joint work generate our consciousness. *"(Dr. Bruce Lipton)*

Naturally, ***the work at your self-consciousness at the cellular level*** is what is needed in the physical dimension of the one on the path of Self-Resurrection. Interestingly, the quality of our cells is also related to the righteousness of our behaviors and the change in the basic substance of life in us – "**THE SOUL'S LIGHT**." (*Dr. Frederick. Bell).*

Dr. Bell writes that this basic light is both positive and negative like everything else in the magnetically charged world. He applies the terms fire and water to the masculine and feminine principles, respectfully. H also write,

"***As evolution advances, feminine souls display more and more of the fiery masculine qualities, while masculine souls take on the ethereal beauty of the feminine***." Our growing physical intelligence makes us understand that **it is pointless to verge a fight against the gay movement**. Gay people are prone to their self-development as anyone else, and their choice is the matter of their own soul light and their own soul evolution. The most important thing is the richness of a person's intelligence and **the level of his / her self-consciousness**.

Ellen DeGeneres, for example, is a sparklingly intelligent, giving, and radiant person. She illuminates the world, and this is what counts most! Regrettably, forces of attitude, hate, judgment, and ignorance prevail as lower levels of intelligence in this world. So, ***self-consciousness needs to be heightened so a person can be good, not just appear to be good***, and respect humanness in everybody, not just appear to be human. **Live and let live!**

Clear Mind + a Cool Spirit + Calm Guts = Connected Consciousness Smarts!

5."Be Immune to the Poison of Life."

Willpower is our inner gravity! Willful self-sculpturing comes out on stage next, and the words of *Carl Yung* above become pivotal in **SELF-ACCULTURATION**. One of the main exercises in developing your willpower is ***to do something that you don't want to do*** or want to put off till tomorrow. You shape your mind and your willpower by doing something unpleasant intentionally.

So, ***shake off the old dry leaves from your Tree of Life***, and let it grow new ones, strong and green! We need to "**will our lives more**" to be able to resist temptation that gets us off the track of self-perfection once the aware attention slacks. Inner integrity, formed by will, is channeling the mind toward consciously ***perceived*** happiness, and it helps us modify ourselves to attain it.

"***Be Immune to the Poison of Life!" Will Your Life More!" That's the Law!*** *(Carl Yung)*"

Willpower is not inherited, it is developed, and it should be instilled in our kids as early as possible. All the kids live in the world of super-heroes, princes, and princesses, and they are all living in the world of unbelievably heroic deeds. They do not doubt their abilities to do the impossible! We generate these doubts for them. I love the concept of the cartoon "The ***Tale of Despereaux"*** , in which the main character, an ugly little mouse, when asked by the Princess if he was a mouse, he answered, "***No, I am a gentleman. I am a man of honor!***" And he acts as one

. We are what we think we are, BUT NOT WHAT WE REALLY ARE!

So, the best image of ourselves should be retained in the brains of our kids and teenagers. Soon, they will be **molding themselves** with the help of their robot friends that will help them discover their ***exceptionality.*** It must be *a bi-directional process*, allowing the forming human minds affect the machine mind so that both parties can form in their neural nets the most noble qualities.

Auto-induction: **In my Life Quest, I manifest My Best!**

Willful transformation will be successful if we continue to zoom in on ***the Ultimate Result Vision*** of ourselves in five life dimensions and **Inspirational Psychology for Self-Ecology**. We cannot change ourselves if we are not inspired inwardly to do so.

Auto-induction: **To Be Inspired, Be Self-Inspiring!**

No one is going to take you by the hand and lead you forward with the light torch of ***Prometheus*** in hand to lighten up your way to higher consciousness. **Inject willpower into your every cell to self-excel!** Use technology for your self-ecology! Even the best of the gurus and spiritual teachers are just showing the way, but it is up to the followers to take it or to step off the **SELF-ACTUALIZATION** track. There is an old Swedish proverb that says

: ***"The best place to find a helping hand is in your mind and at the end of your arm.***"

Your Mind is Your Best Helping Hand!

6. Defy Common-Sense Gravity, Fly!

Mind it, please that **the URV** ***(the Ultimate Result Vision)*** of the best of you should be modified at every stage of your intelligence development, and it should be crowned by an integral self-image **THE VINTAGE YOU** It is crucial to rely on your own **objective judgment of yourself**, your own self-reflection in your mind and heart. You alone know what you are improving and how well / badly you are doing that! Seeing yourself in the mirror of other people's reactions, their perception of your changed behavior will also help if you perceive it objectively**.**

Auto- suggestion: **"Perception is more important than intent!"**

You will zero in on ***the Ultimate Result Vision*** of the best of you all the time and in every stratum of your life-***physical, emotional, mental, spiritual, and universal***. Just get into the habit of a quick **SELF- SCANNING in** five dimensions before you fall asleep, following the paradigm ***-Self-*****Synthesis -Self-Analysis -Self-Synthesis.**

Give yourself grades for each level, scanning your accomplishments / failures that day Finish your self-scanning with the general picture of your action on the path of bettering yourself in small steps.

Most importantly, focus on what you have learnt about yourself and for yourself that day. Physical intelligence is not just about exercising and health, but self-caring. **It is also reasoning about your life** and mindful, conscious attitude to anything you do in it. Interestingly, we always remember those people who added grains of wisdom to our memory bank.

With the URV instilled in your mind consciously, you won't let anyone ever change it or destroy it. ***"Be the thing in itself***!"*(Hegel)* Don't declare your taking any transformational courses, don't boast about any change that is going on in you. That's your personal business, ***your exceptional mission of Self-Installation,*** backed up inwardly with the autosuggestion:

Auto- Induction: **I know who I am and Who I am Not!**

I am a strong, calm, bold, and determined owner of my firm willpower!

I can…; I want to…; and I will…!

I am becoming better at… with each coming day!

This simple, easy to memorize auto-suggestive mind-set, inducted, when necessary, will boost your spirit and shape you in the desired direction. It is programmed solely by you, not by society, your family, friends, or anyone.

In fact, when you think that you have a near perfect picture if you have yourself, modify it at the Mini, Meta, Mezzo, Macro, and Super levels consequentially.

Defy the Commonsense Gravity, Fly!

7. Be Aristocratically Refined in Your Mind!

Being mindful, polite, kind, respectful, and compassionate means being ***aristocratically refined***. You stop being an impulsive, automatic, and unconscious being, you get led by ***the vintage vision of yourself.*** We can draw an analogy here with **the Ariadne thread** from a well-known Greek myth.

The myth tells us that the daughter of King Minus, ***Ariadne***, who fell in love with ***Theseus*** and saved his life with a skein of linen thread that Theseus had fixed at the entrance of the labyrinth of death where he was supposed to fight the monster ***Minotaur*** and die. Thanks to that **Ariadne thread**, he found the way out after his successful battle with the monster Like ***the Ariadne thread,*** we need to visually fix the **end of our mental cord to the Universal *Intelligence*** and then go through the maze of the labyrinth of life , winding it higher and higher, closer and closer to the to the Ultimate destination – a much more conscious you - ***the Vintage You, and the Integral You!***

Self-Induction: **I am having the rebirth of my self-worth!**

Forming the URV in your mind, you should not forget to take care of your energy field that gets easily demagnetized with negative emotional information of anger, fear, worry, envy, etc. Much dirty emotional information is also coming from other people that are charged with negative energy, thoughts, emotions, and moods.

The URV is my universal key! It's my testing point!

You need to install it in your mind for the entire life, choreographing it in each dimension with your self-growth consciously .***We all have goals in life, but those goals are normally very obscure and materialistic – getting a better paying job, buying a house, a new car, a cruise***.

These are wishes and wants that are quite justifiable, but they are not goals. because the person who has the goal of accomplishing something in life does not see himself in that role every day. **The material quality of life should not outshine personal self-perfection** and the necessity to raise self-consciousness in oneself and the kids, if you have any.

Mind-molding and soul-enriching are everyone's full-time job pitching!

This auto-suggestive vision (***the URV)*** of the best of you in your mind. You will know what to do and how to help yourself when you feel weak, confused, or scared. Having the URV of yourself, you will try to learn more, be inquisitive and curious like a kid, with an insatiable hunger for knowledge. The same qualities humanoids should have, enriching their mechanical brains with better designed skills of ***Emotional Diplomacy***, ***Self-Gravity, and Attitude of Gratitude.*** Be sure to instill into the machine mind the mind-set:

Stop Criticizing and Moralizing . Start Wising!

Valid Information Processing + Monitoring Our Best Innermost Qualities = External Behavior Transformation.

8. Self-Gravity Skills and Morality Refills.

It is paramount for us now to polish both sides of the coin, "***the Moon, and the Sixpence***" sides (*Somerset Maugham/ the novel" The Moon and Sixpence*"} with the priority placed on the values of the intellectual development, without making it just the need to get a career. The sixpence side of such education. aimed at just making money is killing the Moon side of many young talents.

The paradigm of Being and Having works only in one direction -.loving and self -refining!

We need to learn **CONSCIOUS THINKING,** which is slow, systematized and strategized It is the ability to see things **from a bird's eye view, holistically.** The rule that helps us survive in the fast turmoil of life is the ability to think and act holistically with the feeling of gravity - **SELF-WORTH** that helps you ground any doubts, fears, negative comments, etc. You know who you are, and you never forget to follow your thoughts and actions in the Synthesis-Analysis-Synthesis conscious way. The **CODE OF THOUGHT IS YOUR FORT.** It is also your strongest forte!

Generalize -Analyze - Internalize - Strategize –Actualize!

People who are worried about displaying their intelligence display their ignorance. The ultimate result is no self-realization due to wrong priorities in life. Let's , but let's hope you will be an exception. "***Do it and then the money comes; not" I'll do it when the money comes."*** , Socrates was teaching his pupils,

"Prefer knowledge to wealth, for the one is transitory, the other is perpetual"

Another aspect to consider is the ***lack of*** **NOBLENESS.** The nobleness which I refer to here originates from **intelligence and stable morality** that are devoid of **unconscious, automatic living**, '***hullabaloo"*** speaking, uncontrolled emotional reactions, unfaithful relationships, *and* irrational behaviors that display the absence of **SELF-GRAVITY SKILLS.** *(the book "Dis-Entanglement")* To beat the fierce competition, generated by AI, we need to grow intellectually, ***developing the basic vistas of intelligence*** in synch with the newest scientific developments in five levels –

Mini, Meta, Mezzo, Macro, *and* ***Super*** *levels./ Physical- emotional, mental, spiritual, and universal)*

Ignorance that according to *Albert Einstein* is" ***the worst enemy of the humanity***" needs to be stopped not by just the influx of money into different social programs, ***but with a substantial influx of intelligence into the global social and educational culture.*** ***"Education is a progressive discovery of our ignorance."*** *(Bill Durant)*

Serving money and prioritizing its value over ***the necessity to raise our self-consciousness***, humanity will still be drowning in wars, bloodshed, crime, and a blind search for pleasure that makes people forget about the main goal in life - their own **personal transformation and full self-actualization.**

Intelligently Strong Global Health is Our Commonwealth!

9. Our Sanity is in Inner Personal Gravity!

Willful self-sculpturing comes out on stage next, and ***moral immunity*** becomes pivotal in SELF-ACCULTURATION, impossible without instilling in oneself a keen sense of **INNER GRAVITY**. One of the main exercises here is to do something that you don't want to do or want to put off till tomorrow. You shape your mind and your willpower by doing something unpleasant intentionally. So, ***shake off the old dry leaves from your Tree of Life***, and let it grow new ones, strong and green! We need to "***will our lives more***" to be able to resist temptation that gets us off the track of self-perfection once the aware attention slacks. Inner integrity, formed by will, is channeling the mind toward conscious installation of a **SELF- GRAVITY SKILL.**

"Be Immune to the Poison of Life!" *(Carl Yung)*

SELF-GRAVITY is not inherited, it is developed, and it should be instilled in our kids as early as possible. All the kids live in the world of super-heroes, princes, and princesses, and they are all living in the world of unbelievably heroic deeds. They do not doubt their abilities to do the impossible! We generate these doubts in them by demonstrating our own lack of gravity skills

I love the concept of the cartoon " ***The Tale of Despereaux"*** , in which the main character, an ugly little mouse, when asked by the Princess if he was a mouse, he answered, " ***No, I am a gentleman. I am a man of honor!***" And he acts as one.

We are what we think we are!

Self-gravity that installs the best image of ourselves should rein in the brains of our kids and teenagers while they are getting transformed into their grown-up lives. The process of this growing is what we need to focus on, ***not just an ephemeral goal that is often dead set.*** Your constant, self-inspiring auto-inductions should be:

In My Life Quest, I Manifest My Best!

Our willful transformation will become more achievable if we continue to zoom in on these words to obtain more strength, willpower, and perseverance. They will help you go through a reflective process of self-assessment of your deeds, *on the one hand*, and ***mindful, purposeful channeling of your intelligence***, on the other.

No one is going to take you by the hand and lead you forward with the light torch of Prometheus in hand to lighten up your way to a higher consciousness. Be a self-guru to accomplish whatever you want to! Even the best of the gurus and spiritual teachers are just showing the way, but it's up to the followers to take it or to step off the **SELF- ACTUALIZATION** track. There is an old *Swedish proverb* that says: **"*The best place to find a helping hand is in your mind and at the end of your arm.*"** Considering the role of intelligence in our life, we can rightfully declare auto-suggestively, "**My mind is my best helping hand!**"

New Knowledge + Digitized Skills = the Know-How of Self-Development!

10. Staying Young is Real Fun!

Physical Intelligence is inseparable from our spiritual health and the holistic development of its human fractal.

Aware attention to the **inner integrity,** consciously monitored by the mind, ***uplifts the body's vibrations,*** changing the frequencies of our thoughts, words, and feelings about our health, and the entire physical state. We heighten our consciousness, strengthen our immune system, ***and rejuvenate the body***

For sure***, life-like machine clones will not have the problem of aging bothering them***. Will they be eternal? In fact, the longevity problem is the one they are helping us implement even now. You can also inspire the mind for a much longer, younger life. ***When we think young, we become younger!*** In fact, there are three different ages that we have:

1. Chronological - ***how old you are*** *according to your passport.*

2. Physiological - *how old* ***you feel*** *you are.*

3. Psychological – *how old* ***you think*** *you are!* Instill in yourself the program for a much younger age or eternal youth and keep saying auto-suggestively the mind-set below. It will help you stay as young as you want. I was inspired to write it having read the book *"Ageless Body, Timeless Mind" by Deepak Chopra* years ago. It is still working!

"A light heart lives long" *(William Shakespeare)*

What kind of person you are now and what you will be in the process of self- creation will result in what kind of person you will be leaving this earth. Keep suggesting a much younger age to your mind with the feelings of gratitude to your body and God for the gift of life that you have

Do it without regrets, blames, grudges, discomfort, and thoughts about a wasted life.

It is in no way wasted if you are holding this book in your hands. Here's **the Auto-Inductive Program for youth** that I say to myself every morning, looking into the mirror Choose the age you feel most comfortable to project to yourself.

I am 27 *(the age you felt your best)* **and not a day more!**

I am as young as ever before!

I am dynamic, as ever.

I am sluggish - Never!

I was, I am, and I will Be Young - Forever!

11. Form the "Vintage You" Vision with an Utmost Precision!

The ***URV(*** *Ultimate Result Vision****)*** of yourself will be changing at each holistic level and at each age of our self-growth, in the same way as our perception of different pieces of classic literature changes with maturing of our intelligence and self-consciousness. Watching thinking humanoids is really a great motivation for us to become better and much more intelligent.

Interaction with humanoids teaches us the depth of self-perception

We should stop killing time mindlessly, reading junk books, watching stereotyped movies, time entertaining uninteresting people and become more **SELF-REFLECTIVE,** enjoying the works of art that are always an insightful self-expression of the mind and the heart of the author. SELF-ASSESSMENT in five life strata, conducted daily before falling asleep will help you develop an objective vision of Self.

We should seek wisdom in books that enrich us with added information and in people around us who inspire us with their self-growth, not just with the amount of money they make. Even though money needs brains, it is not always that money-making ways raise our self-consciousness. *(See Financial Intelligence below.)*

The growth in ***Self-Awareness***, in intelligence, human qualities, and life-processed wisdom is the best motivator. Such people that have it shine from inside with the **VINTAGE IMAGE** of themselves reformed ***physically, emotionally, mentally, spiritually, and universally.*** We get mesmerized, listening to their insightful revelations. Develop your own uniqueness in these levels, without comparing or contrasting it with someone else's.

Do not compare and do not compete. Be in a unique outfit here and there!

Your growing self-image will fill you up with a profound sense of magnetic power, and it will enrich ***you with personal magnetism***. Your ***Physical Intelligence*** will be transformed into **LIVING INTELLIGENCE. What a great feeling that is!**

However, the most challenging part of self-work at your URV is your ability ***to practice what I preach here holistically*** through studying, practicing, intense reading, thinking, creative imagination, constant spiritual growth, and, most importantly, consistent action, fortified by a firm will! Why? *Holistic growth is a new paradigm in the scientific belief system!*

Every branch of science is developing now toward **the *unification of knowledge and its synchronization with the fundamental laws of cosmos***. Science, technology, business, everything is getting integrated to obtain a more holistically advanced shape, as it had been in ancient times in the **SPARTAN EDUCATION** system in Greece. **LOVE YOURSELF THE WAY YOU HAVE BECOME! Know who you are and who you are not. That's your Personal Fort!**

The Best of You is the Vintage You!

12. "Be Doggedly Tenacious on This Path!" *(T. D. Jakes)*

The words above belong to a very authentic man, the senior pastor at the ***non-denominational megachurch T. D. Jakes.*** I love his book ***"Reposition Yourself!*** The ethical qualities that we need to develop in us and AIs to " **reposition ourselves"** in our perception of **religion+ science+ AIs** should be in the ethical trinity: **SACREDNESS + NOBLENESS+ LOVE=**

A NEW, MULTI-DIMENSIONAL *(Non-Denominational)* **YOU!**

We are One under the Common Sun!

I Am My Best Friend. I Am My Beginning and My End!

1. Physical Level - *mindful body awareness, health-conscious, nutrition, stress management, language rationality, speech immunity to foul language or negative talking*. Remember, ***you are best friend; you are your beginning and your end!***

Self- Induction: ***Eat less, talk less, fight less, and spend less!*** **Conduct a self-restriction war - less is more!**

2. Emotional Level – emotional awareness, c*ontrolling emotions and* developing *the feelings of love and synergy.* Your main induction should be:

Self-Induction: T**he smile, the posture, and a good mood are my emotional food!**

3. Mental Level - mental awareness - a *lot of quality reading, thinking, visualizing, rationalizing, planning, systematizing, studying, and positively projecting, professional realization, creativity, being truly mindful at any action that you perform.*

Self-Induction: ***"Fertilize your mind with thinking!"*** *(Albert Einstein)*

4. Spiritual Level – ***Spiritual awareness****, honest service, interpersonal development, cultural enrichment and responsibility to the society, commitment to self-refinement,* grea*t values, sincere faith.* Your main induction should be:

Self-Induction: ***Make your mind kind and your heart smart!***

5. Universal Level - ***Internalized rules of living***, develop ***spiritualized intelligence***, *correct behavior, unified thinking, elevated level of consciousness, new horizons of perception of entire life. Your main induction should be:*

Self-Induction: ***I am One with Everything under the Sun!***

Tame Your Analytical Thinking to Give the Visual, Systemic Thinking a Chance!

13. The Sense of Measure is a Great Treasure!

In sum, ***Physical Intelligence*** is the basic one, and it should be enhanced with **the SENSE OF MEASURE** in our perceiving the life's pleasure and the price we pay for it if we disregard this principle primarily!

Higher physical fitness generates higher mental fitness! *(And vice versa)*

So, will-power, self-discipline, self-control, and the ability to deny oneself the excessiveness of any sort: harmful foods, drinks, extra clothing, accumulating different miscellaneous items, pointless talking, tongue-lashing, impulsive relationships, uncontrolled sex-drive. meaningless texting or calling someone need to be cut in the bud!

Self- Induction: **Less is more! That is the Law!**

Remember, change happens in a spiral upward way - two twists up and one twists back.

Sometimes, two twists happen to block the advancement. So, you ***need to condition yourself auto-suggestively for consistency*** and go through processing your self-creation with a sane head. True physical intelligence is still working when your mind and the spirit, not the doctors, are involved, "***God cures, and the doctor sends the bill.***" *(Mark Twain)*

Keep intellectualizing your heart and emotionalizing the mind by magnetizing your inner core.

Do not get inflated with the crowd mentality of common-sense **THINK FOR YOURSELF!** Magnetize your center with the mental power *(Cognitive Intelligence below*) to develop a strong personal core to which people will gravitate for guidance, inspiration, and inner strength.

The Ripping Effect of Kindness is Never Mindless!

Finally, ***be unique, free-spirited, and magnetically attractive*!** Belonging to any social or religious dogma, or chauvinism of any sort did not make humanity any better or any happier, but ***personal richness*** has inspired and uplifted people for centuries.

According to *Albert Einstein*, **"Everything, even lunacy is mass produced**." For the sense of measure to get better rooted in you, induct yourself with:

Live a more aware life! Be alive!

Our life equation of pressure and pleasure

Comes in the bits of treasure

That only God can measure!

The Art of Living is the Art of Self-Learning!

Cognition Transforms You from a Cocoon to a Butterfly!

Don't Crawl – Fly!

he

(Design by Yolanta Lensky)

The Five Elements of Life – Space, Fire, Air, Water, and Earth Determine Our Developmental Growth!

(Universal, Spiritual, Mental, Emotional, Physical)

Stage One, Division 2

General Intelligence, ***Physical Level*** / *Self-Awareness*

Cognitive

Intelligence

"Don't teach a thought. Teach to think!"

(Immanuel Kant)

" Don't teach just a subject. Teach the whole person!

(Leo Vygotsky)

Cognition is Holistic Knowledge without the Left - Right Brain's Division!

The Intellectual Odyssey of a Seeking Soul

Vistas of Intelligence to install next:

Let AI Enforce Technological OutBurst!

10. Universal Intelligence — **Self-Salvation**
9. Spiritual Intelligence — *(Universal Realm)*

8. Social Intelligence — **Self-Realization**
7. Cultural Intelligence — ***(Spiritual realm)***

6.Financial Intelligence — **Self-Installation**
5. Professional /Creative/Intelligence — *(Mental Realm)*

4. Psychological Intelligence — **Soul-Refining**
3.Emotional Intelligence — *(Emotional Realm)*

2. Language Intelligence — **Self-Awareness**
1.General Intelligence — *(Physical Realm)*

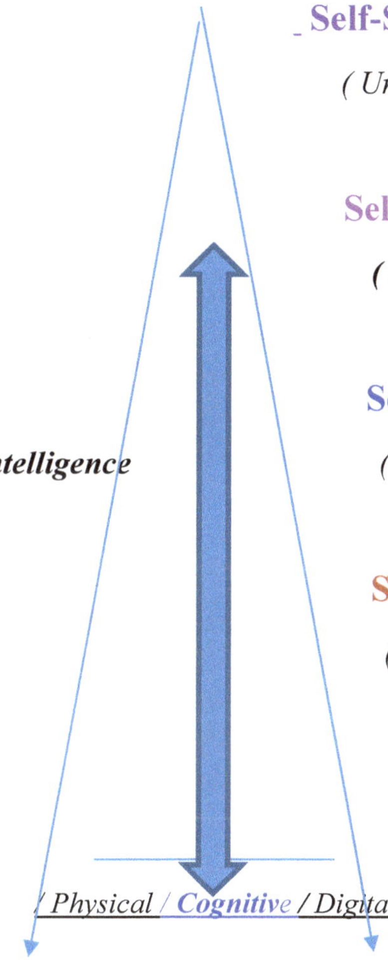

Fertilize Your Mind with a Digitized Brain Rewind!

To Make Your Life Digitally Wise,

Generalize - Internalize - Personalize - Strategize –Actualize!

Don't Be Mind-Negligent. Be Mind-Intelligent!

1. Cognitive Intelligence is Our Primary Self-Genesis!

The second vital part of General ***Intelligence*** formation is **COGNITION-** **the necessity to know more** which is, in fact, the essence of human existence. It is not just intelligence; it is the skill of survival and the essence of **SELF-CONSCIOUSNESS** formation.

A new approach to knowledge accumulation, or ***cognition,*** is required at our time not only to sustain the pressure of the exponential growth of technology and a fierce competition, generated by it, but also, ***to deal with the inner pressure for the necessity of a personal self-expression*** that keeps us wishing to erupt with the force we had never known so far.

Cognition is also our human developmental mission!

I see ***Cognitive Intelligence*** as a sub-level of the physical one because **language and cognition are the foundation for our human mission.** I started the holistic paradigm of intelligence formation with ***physical intelligence*** only because it has overwhelmed humanity. We are more conscious of the physical aspect of survival now, attending the gyms and prioritizing the fight for weight loss and our muscle tone. The body is the most intelligent machine that accumulates the memory or the intelligence for us to function at the biological and mental levels in an inseparable unity. ***But the sense of measure that I have mentioned above is the essential stuff!***

Body + Spirit + Mind + Self-Consciousness + Universal Consciousness

I keep accentuating an evident fact that ***intelligence enrichment is getting increasingly neglected by our youth.*** Pleasure mongering, immediate gratification of it, money-chasing, drugs-addiction, and alcohol are consuming many young, gifted minds, and this book is written for them, as a wake-up call. *Our intelligence is closely interwoven with our consciousness that is governed by Universal Intelligence*

Meanwhile, **"*"Ignorance is still the worst enemy of the humanity*.**" *(Albert Einstein).* How great and how simple these words are and how often they are repeated to no avail! ***No technology can heal ignorance*; *we need to eradicate it ourselves*** by discovering new and new dimensions of the unknown.

"The only true wisdom is in knowing you know nothing." *(Socrates)*

Universal knowledge is infinite, but our personal knowledge of everything should not be ***finite*** because it is a never-ending process of the evolution of life.

Fortunately, we are gradually realizing an urgent need for ***converting the automatic processes of digitizing our minds into the conscious processes of controlling and re-programming them*** for the goals of our evolution. **Feel but think! No Brains, No Gains!**

"We must use the brain to intentionally alter the way in which it is organized and the way it makes connections to rationalize our behavior." *(Deepak Chopra)*

The Power of Cognition is Mind's Knowledge Ignition!

2. Knowledge is Chaotically Misleading Now

I have outlined **Physical *Intelligence*** as the first stage in the holistic paradigm of the system because it initiates ***the process of self-quest*** which is people's life commitment. now. We are much more aware of the physical state of the body, presented by medical science ***as a system,*** now. It is a great evolutionary accomplishment, but we must yet realize that ***our bodies are run by the highest intelligence*** and this intelligence is the product of our evolving **self-consciousness**.

We do not run our bodies; the energy of Super Consciousness does!

There is no intelligence without knowledge, and "***the world is the mountain of knowledge***." *(Albert Einstein)* There is no intelligence without cognitive intelligence because **cognition means knowledge**. That's why Cognitive intelligence is at the base of the intelligence paradigm. It is intelligence itself. Cognition is the process that requires ***"cognitive structuring and restructuring of our lives.*** *" (Watson and Tharp)* Digital Intelligence is at the bottom of the holistic pyramid, too, because it is a vital mind-forming part of the indigo children's population.

The book just outlines the major points of intelligence to master because self-consciousness is based on intelligence, and it is exceptionally low with us now because it is not holistically based. ***"Science proves that mind-body connection, blood vessels, brain cells, and cognition are related. A conscious mind becomes activated through the accumulation of knowledge and overall awareness of one's place in the river of life."*** *(Rave. .PS. Berg)*

Self-awareness generates the inner force with the help of growing self-worth!

The thinking process converts knowledge into energy, and ***it is mental energy that is our engine in life***. Science teaches us that the structure of matter is dependent on Universal Intelligence or consciousness that is still the puzzle for us to unravel and that we are deciphering with the help of Artificial Intelligence now.

It means that ***the acquisition of knowledge is not merely a transfer of information from one source to another.***

The roots of knowledge run far deeper than the intercourse between a mother and a child, a teacher and a student, a body and the mind, a computer, and a user. **Elevated Consciousness is our goal!** It is our responsibility to change our cerebral script and elevate ourselves consciously in five dimensions integrally.

There are many people that can be called "***giants of intellect, but pigmies of consciousness."*** *(David Eike)***.** Knowledge is not just accumulating information that makes us more powerful.

Knowledge becomes power only when it is processed consciously, selected for its validity, and is turned into real wisdom, having been verified by the right action.

Those that Defy the Gravity of the Common – Fly! Those that Crawl – Die!

3. Tips to Jump-Start Constructive Thinking:

Thinking is the process of operating the mind in synch with the brain, synchronically and consciously. As science has it , ***"the quality of our thinking depends on further development of the right hemisphere,"*** and, therefore, thinking also needs to be enhanced with knowledge about the brain and self-managing skills ,essential to develop ***Professional Intelligence*** that I am exploring below!

Your professional and personal mission must be grounded by your personal philosophy that must permeate your entire being, generating your personal integrity.

"I cannot teach anybody anything. I can only make them think. *"(Socrates)*

1. *Universal Mind choreographs everything that is happening in our lives*. Our talking to ourselves while meditating is, in fact, self-programming every cell in our body, every mini-computer, uploading them with our daily programs.

Our thinking also must *work in sync with Master Mind the Universe!*

It means that we must learn to upload our brains consciously to construct our thinking that is dissipated, unmanageable, and messy otherwise.

Be aware of your thoughts! Don't let them turn into UNCONTROLLED *warts!*

2. *"There is no true thinking without discipline."* *(John Miller)* Discipline requires thinking slowly in a step- by-step way, zooming on the messages that the mind gets. Thinking slowly does not mean that you are dumb. Thinking slowly makes it possible to sort out the incoming information **in a conceptually governed way** into special compartments of sorted out knowledge.

3. *Being logical is another demand*. When we are logical, we can organize the information into ***chunks*** and store it carefully in the brain. The help of AIs here cannot be underestimated.

Compartmentalization of information is an organized brain formation!

Storing information in the window-frame way is essential for constructive, expansive, and developmental thinking skills that we need to rationalize our life. Interestingly, the electrical circuit in the brain lights up when we speak or write logically.

Think on paper to build up logic better in speaking!

4. *Build a language-control muscle*! *(See Language Intelligence below)* Thinking and language are inseparable in the brain. Writing develops strategic thinking and a strategy-oriented mind. Speaking creates an emotional personality and listening shapes respectful man.

"Don't Say Everything you know but Know Everything You Say!"

Rationalization of Thinking is a Must!

4. "I Am No Einstein. I'm Just a Thinker." *(Albert Einstein)*

"An intellectual has ready answers for everything; his mind is closed; it's frozen in his patterns of thinking. He is knowledgeable, but not intelligent. " ("*Intelligence, "Osho)* We should not live on borrowed information; we should seek the information that expands our own intelligence, independent of the opinions of those individuals whose intelligence is damaged by "***collective psyche*** or ***the collective unconscious***" *(Carl Yung).*

"Real intelligence is an open sky. No one limits you!"*(Osho)*

Albert Einstein never identified himself with the image that the collective mind had created of him. He remained independently authentic in his mind, in his "***thinking laboratory.***" Long after he became an icon, a celebrity genius, he ruefully and most modestly confessed,

" I do not care about the details. All I what to know what God thinks!

A holistic mode of knowledge acquisition that this book is advocating for is not a mechanical combination of thought patterns. ***It is pre-programmed by the whole evolution. We*** cannot but marvel at the honeybees that use abstract reasoning, learning, accurate decision-making and display an amazing amount of intelligence. But our learning should be very selective and time relevant.

There should not be any chaotic, mindless rout to learning anymore. **INFORMATION PRESENTATION** and **INFORMATION PROCESSING** must be well-sorted out for the validity of the data for our ***AI focused intelligence***.

To cleanse the memory banks and upload them only with the well-selected, time-relevant, and science-verified information, we need to instill new thinking skills in our learners. *(see below)*

Researchers observe a certain amount of intelligence in plants and even in stones. Paying ***aware attention*** to nature and to life in its every form, we are learning to become not only life-conscious and life-appreciative, but also more attentive to the way we learn things, observing life around us and following its advice. Auto-Induction:

I have spiritual glee.

I am in love with life, and life is in love with me!

Keep Working with your Self-Inductive Talking!

(Upload your smartphone with a new digital tone!)

In My Thought, I Plug Myself to the Universal Life Volt!

5. The Art of Thinking is the Art of Conscious Learning!

It is extremely hard to live, not to think. However, the greatest part of people's thinking is involuntary, automatic, repetitive, and therefore, it is devoid of real quality that conscious thinking should manifest

Since the mind is conditioned by the past, we are pushed to repeat the past habitual actions repeatedly, governed by the subconscious mind that is not yet holistically rewired. The duality of our perception of the world and automatic thinking results in automatic learning that results in ***accumulating unprocessed knowledge*** that only clogs the brain. Ignorance manifests itself in our automatic actions that we always find the way to justify. In fact, self-justification for the repeated mistakes stands in the way of self-modification and self-actualization, or in the way of the ***rationalization of thinking through the paths of logic, with the aware attention guiding the process.***

A creature of habit is unable to reason,

Make correct conclusions and take a reasoned-out action because he /she is ***lacking in thinking culture***, to begin with. Lack of reading means lack thinking resulting from the lack of knowledge. ***"Most people would sooner die than think; in fact, they do so."*** *(Bertrand Russell)*

Mind without conscious control and order becomes neurotic, unbalanced, destructive, and restless "***One of the greatest blessings in the world is to have sound, healthy mindfulness. The one who possesses it is the most fortunate. When you achieve such healthy mindedness, you are a free human being!"*** *(Delia Lama)* But this mindfulness comes only with a self-educated mind. Healthy mindfulness means having a mind devoid of inner conflicts, uneven emotional reactions, and obsessive behaviors. **PERSONAL INTEGRITY** is the job of an educated mind.

Even though thinking comes naturally to us, ***we need to study the techniques of productive, self-monitored, and self-constructive thinking.*** The process of uploading the information into the biological computers, our brains, needs to be modified , cutting the automatic, uncontrolled perception of it in the bud.

We are of a four-fold bind – the soul, the body, the spirit, and the mind!

The mind unites these four parts into one - the **ULTIMATE RESULT YOU, VINTAGE YOU** or **INTEGRAL YOU*!*** Science shows that we are of two minds. The left hemisphere is rational, analytical, and logical. ***The right hemisphere is nonlinear and instinctive***. The left hemisphere reasons sequentially and excels at analysis. The right hemisphere recognizes patters holistically, creates, and interprets emotions.

"It's most vital for us to develop the right hemisphere now because it is responsible for our ability to comprehend life conceptually and consciously." *(Daniel H. Pink)*

The Art of Living is, in fact, the Art of Mind-Control!

6. Mental Architecture Must Be Holistic in Structure!

Mental architecture is the core of a thought process. *"**New information that is flooding our brains competes with the old in-put, and it gets processed only if the information-processing system is in order." (Scientific American Mind,"** May / June 2014)* We are the ones that are supposed to consciously monitor this process with AI's help now.

Then, like a good intelligence agency, information is stored in the brain and shared only on a need-to-know basis, ***like a programmed humanoid does.*** Rationalizing everything in our lives and, therefore, building up the **AWARENESS of the CAUSE-EFFECT MODALITY**. It requires that we develop our critical thinking skills at every level of intellectual self-development and self-actualization, not allowing automatic reactions to occur on an automatic, sub-conscious basis.

Putting the mind in charge of our emotions and actions means rationalization of each step we take. Gradually, you will be monitoring your physical, ***emotional, mental, spiritual, and universal strata,*** gradually becoming a self-made person with a determined will.

Auto-Induction: ***I think consciously, I speak consciously, I act consciously,***

I breathe consciously, I eat consciously, I drive consciously.

I'm a New Me; I'm a Conscious Me!

In fact, any action you take, however small, needs ***to be consciously monitored*** by the aware attention and the information that you operate at that moment. The input of information needs to be digested mentally and structured consciously **on *five basic philosophical levels.***

Probing the information for its rationalization

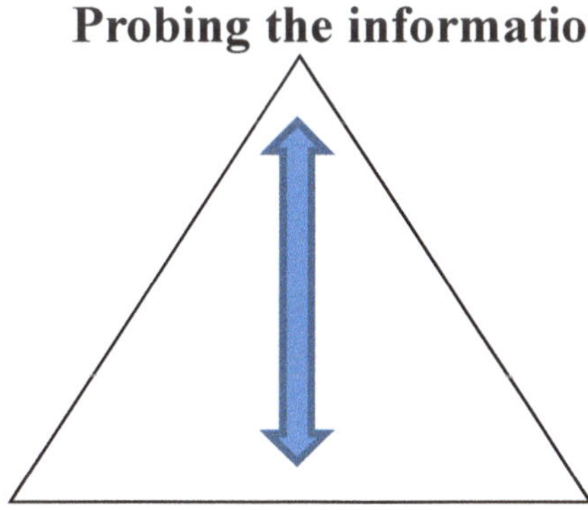

Super **Level**	**Reasoning**
Macro Level	**Generalizing**
Mezzo Level	**Conceptualizing**
Meta Level	**Analyzing**
Mini Level	**Contextualizing**

Business -thinking ais a very constructive thing, and the best thinkers make the best money!

Another conscious act is our dealing with the information is its careful sifting for its validity for your professional growth, general intelligence, life problems, love preservation, etc. We must prove any piece of information for its validity and importance for our self-growth at every level of our general intelligence and through five rationalization steps, presented above

Generalize -Analyze - Internalize - Strategize – Actualize! Be Wise!

7. Reign Over Your Brain!

Integral mind-formation requires substantial development in every level of intelligence at least in a dilettante way and ***a true modifying of your subconscious mind*** that needs to be consciously channeling the brain to the ultimate result - a consequential rise of self- consciousness that will let you re-program your brain by altering its working memory. Your biological computer will get re-charged and re-programmed.

"You are here to enable the divine purpose of the universe to unfold.

That is how important you are!" *(Eckhart Tolle)*

Our personal knowledge is the structure of everything we believe in. Every concept, opinion, belief, and value form a little branch of **the Tree of Knowledge** that each of us needs to grow as a branch of _ that we will leave at some point as of one of its precious leaves.

Psychology proves to us that the feelings of self-accomplishment give us the greatest elation and true feelings of happiness. *The main character of a great movie "**The Pursuit of Happiness"** with an excellent **Will Smith,** remembers how happy he was when, as a school student, he got particularly good grades, anticipating great accomplishments in the future.*

*However, his dream never got realized. Finally, having gone through all kinds of troubles and tribulation and having upgraded his intelligence and done a lot of life-learning, he proved to himself that **he was tougher than his very tough life.** He became a first-class broker and was offered an excellent job.* This is when he realized what the pursuit of happiness was all about. He said,

"This is the day of my true happiness. I have created it myself!"

In sum, the book "***Digital Binary and Intelligence Refinery***" is, in fact my attempt to persuade any girl / boy, of any race, religious faith, absence of prior education and lack of money to get solid, world-recognized one to remember that **our brains have no color**, and our abilities to expand them are boundless for every mind. The world is erupting with technological geniuses in every corner of the world. The time for **HUMN EXCEPTIONALITY** has come! *(See the book " **"Exceptionality"** / Emotional level/ Digital Psychology for Self-Ecology).* That's why the title of the first book, featuring the initial, physical level of self-creation in the ***System of Self-Resurrection*** *(See Book Structure above)* ***is:***

I Am Free to Be the Best of Me!

The book provide a comprehensive frame of reference for developing holistic intelligence as the main prerequisite for a **professional and personal Self-Installation in life**.

You Don't Need to Justify Yourself. Know Who You Are in Every Cell!

8. Digitized Character Filling Secures Inner Stilling!

Present- day AI technology creates robots with distinctive character features, but we need robots that will correct our bad traits of character that pollute our thinking and turn into chaos our emotional and psychological. Intelligence.

The lucrative side of the business of humanoid designing and producing them for entertainment purposes should be secondary in our evolutionary agenda.

I have noted in the **Introduction** that the main goal of our revolutionary upheaval, connected with the unprecedented expansion of *Artificial Intelligence* unleashed the amazing creativity of the brain, boosted the communication techniques, **connecting both brains** by putting the left and right hemispheres in sync. However, it has not produced so far anything of the fundamental importance for our evolutionary growth - **our hearts and mind are not in synch!**

We think, but we do not feel, and we feel, but we do not think.

I am sure that the creation of the specially designed robots , in the form of a watch, a, or any other device , able to perceive and monitor the vibration of our emotional glands ***Amygdala*** could be used to monitor our fears, impulsivity, irritability, discontent, tongue lashing, etc.

Watching humanoids during different interviews , we cannot but notice how well they manage their emotional framework, and this is what we need to learn with their vigilant monitoring our emotions in our neurological connection .

It does not have to be a chip in the brain . It may be a wristwatch or a finger ring that will be handy in the situation when we may lose our cool and control.

Timely warning us ***to pay attention*** to a mechanical, impulsive action or a negative emotion generated desire to disregard **EMOTIONAL DIPLOMACY** rules will turn on the conscious response and our unhealthy habits will be stopped at will.

"***Conscious awareness of our feelings will help control them at once***." *(Dr. Dispenza)*

As *Albert Einstein* most insightfully noted,

"Bad habits have a good tendency- either you kill them, or they kill you!"

Artificial Intelligence is successfully expanding our intelligence through the brain gam es variety, the avalanche of information on our digital gadgets, and though online social media.

Amazingly, humanoids, able to enlighten our communication in science and other areas are being produced by different companies now, but we also need the devices that are focused on heightening our **SELF-CONSCIOUSNESS** and making us more loving, caring, considerate, compassionate, understanding, tolerant, patient, and truly **GODLY** *de facto*, not just *de juror.*

It is Hard to be Godly in a Godless world!

9. Let us Strengthen Our Human Gene with the Digitized Mind's Hygiene!

You Are How You Think!

Super Consciousness

Super-Conscious Mind

(Concluding Synthesis*)*

Spiritualized Intelligence

Intuition /telepathy in action

Having intuitive insights

Generalizing *(inferring /concluding)*

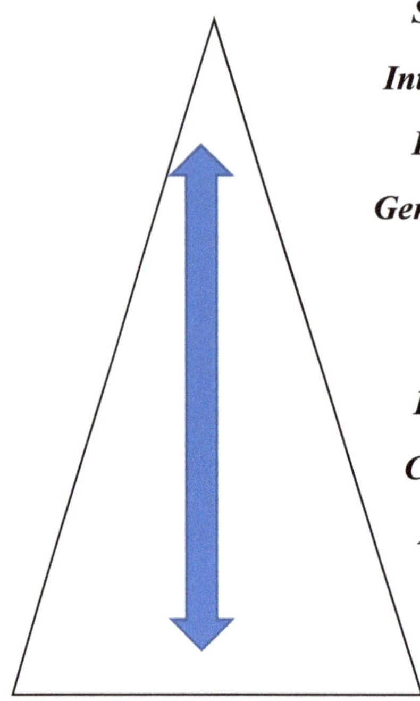

Conscious Mind

(Analysis)

Rationalizing / Reasoning

Inductive / Deductive *thinking)*

Conceptualizing / Creating

Analyzing / Justifying

Subconscious Mind

(Initial Synthesis

Visualizing / Perceiving

Information processing

Auto- Induction: **I live consciously! I have a thinking life! I'm alive!**

Knowledge has many degrees, and to reach the highest degree of conscious development of logic and all the sophisticated forms of reasoning, or" **pure reason**" by Kant. The framework of the **MENTAL SKILLS** ***presented above*** must be mastered continuously and consciously.

Super Artificial Intelligence is based on **AWARE ATTENTION** ***without negligence***

Teaching these thinking skills is essential in holistic education that I am advocating for here. Then **knowledge becomes true awareness** that, in turn, stimulates **the** development of a person's intuition and insightful, *"**spiritualized intelligence**."(John Baines)* **that we need to instill in humanoids now.**

They are becoming self-aware, but not soul noble!

Rationalization of thinking and feeling in sync is the name of the technique on the path of the AI enhanced our full Self-Actualization Hence, the **job interviewing techniques** should also be holistically monitored to check a job seekers general outlook in five dimensions'

Holistic Awareness of Reality is a Must!

10. Reflective Thinking is Conscience Shaping!

Reflective thinking is inseparable from conscience! The idea of living at peace with conscience is instilled by the parents in any culture because it is essential in all sacred books. It is also at the core of a continuous process of self-construction and consciousness rising. It is also the cause-effect incentive for our prayers, meditation, and self-reflection. It is the starting point for self-awareness, self- judgment, and self-balancing. It provides food for self-inductive work mentally and self-perfecting work emotionally in the Now, not retrospectively as we often tend to do.

"Thinking that is not rooted in the Now is dysfunctional!"*(Eckhart Tolle)*

Every one of us needs to clear up our minds **IN SYNCH** with the heart and make oneself more aware of what is going on in them at any given moment in life. Neuroscience provides us with invaluable information that self-reflection is indispensable in the process of rationalization of thinking and empowering of our emotions.

Self-awareness is impossible without such on-going practice.

That is why it is so good to observe oneself during meditation, a passive or an active one. We can do reflective thinking and actual re-unification with our intuitive ***conscience - scanning*** while walking, running, driving, while putting everything in order inside the house, the office, and inside ourselves, and especially before going to bed, especially if we feel distraught emotionally.

No wonder, women that are constantly putting things in order have a much more developed intuition than men, and they are more **CONSCIENCE- CONSCIOUS** and prone to ask for forgiveness. It is noteworthy to mention again that when we have basic knowledge about how the brain operates, at least in general terms, we can better control both the brain and emotions through self-reflection and auto-induction.

"The power of reflection fortified by the most primary idea how the brain operates will help get better insights into our own thoughts." *("Scientific American Mind," Sep/ Oct .2014)* Reflective thinking should be trained holistically - at every level of intelligence development: **Mini, Meta, Mezzo, Macro, and Super** because your intelligence is directly reflected in your consciousness.

Body + Spirit + Mind + Self-Consciousness + Universal Consciousness

Reflecting on what is bothering you physically, emotionally, psychologically, mentally, and spiritually does not necessarily mean stopping the chatter box in the brain as many books suggest. It is also your active and constant rationalization of everything you think, feel, say, and do while processing it through the **GRID OF CONSCIENCE**

You have a chance to put yourself in order in every dimension though holistic self-scanning.

Intuitive Conscience Scanning is Soul-Refining!

11. Thinking Hygiene is on the Scene!

1.To begin with, both forms of communication need ***constant self-control*** to maintain the **THINKING HYGIENE** to process and actualize incoming information correctly; we need to learn to use aware attention as our scanning tool. I consistently call it **AWARE ATTENTION** because aware attention is the attention that is loaded with conscious perception of the information that had already been sorted out by the brain for its validity based on the prior ***conscious compartmentalization*** of the processed knowledge.

2. Next, **the skill for speed-reading comes to the stage.** The manner of writing scientific articles and other important pieces of information in mass media that are supposed to provide something to consider are too wordy and too professionally specified. The readings meant for popularization and mental digestion, especially by our young, very-digitized minds is often, if not always, beyond their comprehension.

Teaching speed-reading is rarely provided, but this is the skill we desperately need to be able to scan the information for its validity. I always start any course by teaching my learners to speed-read, to skip and scan the information for its conceptual value, connecting this skill with teaching concise writing and speaking.

3. Store the information in **the pre-frontal lobes of the brain.** Science proves that most thinking is done in the perception area - the area that assesses information for its importance. Only with aware attention to our thoughts and words are we able to follow another rule

: **Say what you mean and mean what you say!**

4. Combine your emotions with reason (mind+ heart in sync) with the help of aware attention, paid to everything and everyone around you. Having observed the situation and those involved in it, give yourself a minute to think about your response, **halt the immediate, impulsive reaction**. ***"If you don't have something nice to say, don't say anything at all!"***

Don't verbalize everything that comes to mind! Be One of a Kind!

5. **Since we delegate increased processing to computers, perceptual thinking** that is based on the aware attention habit and ***independent thinking*** becomes increasingly crucial nowadays because it remains to be human.

Finally, **it's crucial to follow every chunk of information that you want to digest with the question:** ***"What do I make of it?"***

Aware attention and perceptive thinking will help you remove the chaos in the mind or clean up your memory storage from the junk that you have accumulated there

,Uplift your spirit and make any job that you are doing meaningful.

Your Aware Attention Must Never Be in Retention!

12. To Be Interesting, Get Interested!

Note it, please, knowing a lot does not make you more intelligent, but knowing what you need to be wise is creating your life's paradise. Without a mindful enrichment of knowledge, **no selection in the brain is possible** because ***there are no points of reference for such mental operations***.

Thinking should start with a philosophizing and general intellectualizing of the problem first, analyzing, or probing the options and arguments of its solution next, and finishing with ***inferring on the outcome***, or by the route:

General – Specific – General / Synthesis – Analysis – Synthesis.

The mind should always ***be consciously following this route,*** in full awareness of its rational necessity while *studying, learning, listening, talking, writing, and working*, irrespective of the field of knowledge or the trade that you operate in. Our school kids should ***acquire critical thinking skills*** not through all kinds of tests, but by way of demonstrating their reasoning ability verbally, not just in a written form.

The background knowledge is the start of establishing* MIND-TO-MIND *and* HEART-TO-HEART *connection with another person.

It's only natural that drugs, as the fun-glazing aspect of life, destroy the brain, attention, and its ability to focus for the needed rationalization of life. Unfortunately, this most negative aspect of life in s not professionally installed in the young people living away from their parents in dormitories of different colleges. It became the matter of course to have alcohol or to smoke pot. So, ***the mind-set of Self-Awareness*** must always be at hand in your mind-depository.

Auto -induction: **I know who I am, and who I am Not! That's my inner fort!**

Selecting the information for its validity and organizing it into meaningful chunks conceptually in the order presented above is a special skill that helps ***establish a mind-to-mind and heart-to heart connection*** in communication. Therefore, ***education should provide knowledge holistically!*** Forming a general idea in many fields of knowledge works much better for young minds than the specialized education with a selective background. This situation widens **the area of intellectual vulnerability of the students** that are not properly informed in some areas of knowledge that they had not chosen to study.

, say. It just was not the subject that they had chosen to master in high school. Obviously, the widening of a general outlook gives a person more something to consider and adds confidence in building up his / her "**NETWORK INTELLIGENCE**" *(Reid Hoffman)* that we are focusing our socializing on these days. ***Intellectual networking is at the core of our evolutionary development now! Don't kill your time with whoever mindlessly.***

Be Alert to Knowledge Gaining and Intelligence Failing!

(End of Stage One - Cognitive Intelligence -Division 2)

Our Life is Now Set on the World-Web Net!

(Picture by Yolanta Lensky)

To Life-Excel, Choreograph AIs, and Yourself!

Stage One, Division 3

*Mini-Level of Intelligence - **Physical Dimension**- Self - Awareness*

Digital

Intelligence

The Whole Brain is Our Technological Domain!

The Intellectual Odyssey of a Seeking Soul

Vistas of Intelligence to install next:

Let AI Enforce Technological OutBurst!

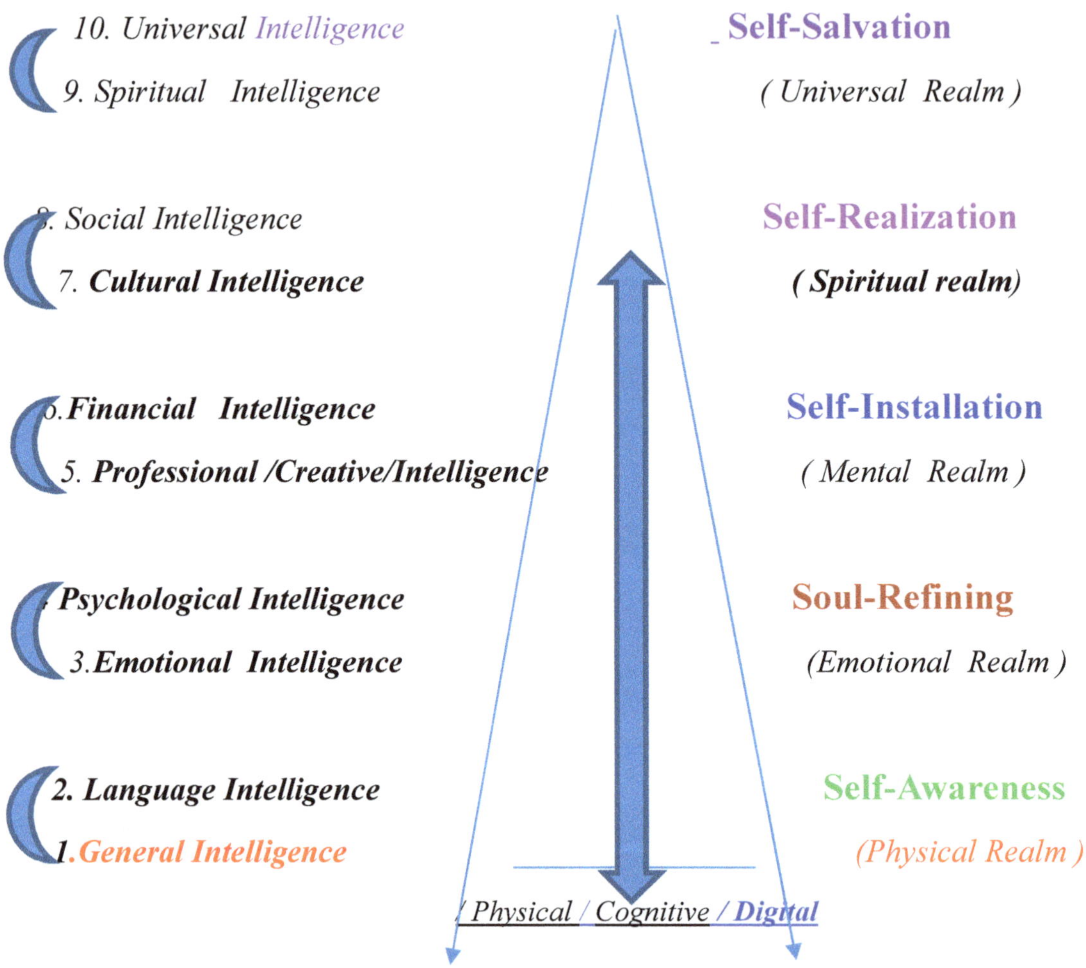

Fertilize Your Mind with a Digitized Brain Rewind!

To Make Your Life Digitally Wise,

Generalize - Internalize - Personalize - Strategize –Actualize!

Don’t Be Digitally Negligent. Be Digitally Intelligent!

1. The Age of Quantum Computing Has Come!

As I have mentioned above, the age of the unprecedented power of the artificial mind is coming to our life stage, and, therefore, considerable improving of our thinking abilities of ***reasoning, conceptualizing, analyzing, and generalizing*** become exceedingly important. These abilities should continue being the prerogative of our general education and self-education. now. Any most advanced computer is a great tool, but it is not Universal Intelligence itself

Undoubtedly, developing **AWARE ATTENTION** to what is going on in life in terms of mechanization of our thinking comes to the forefront of our digital education. Zillions of websites are competing for our aware attention that needs to be channeled with much more ***conscious selection*** on our part. ***Aware attention development is at the core of digital intelligence now***!

We need to adjust our brains to the ever-expanding modes of communication to deal with the avalanche of information and the brutal job competition that electronic evolution has generated. Self–suggestive work is the best reminder to you about your being the owner of your min**d**. It helps de-atomize our thinking." ***Be Conscious. Consciousness mobilizes.*** " *(Neil, Donald Walsch)*

Auto-Induction: **I am becoming more intellectualized, not robotized!**

In sum, the **5** G computing is meant to help our consciousness get enriched ***with higher overtones of intellectual interaction***, based on enriched electronically programmed language intelligence. Digital intelligence is a great part of our intelligence when it is ***consciously monitored by the mind*** in the process of self-actualization aimed at the evolution of humanity in five levels consistently. ***Heighten your magnetic electricity volt!*** See how it integrates into your *physical, emotional, mental, spiritual, and universal* world! Amazingly, with digital intelligence in progress, we are tuning ourselves up to Universal *Intelligence* electronically and physically in some inseparable link, **synchronizing our terrestrial existence with the extra-terrestrial one**

Every Human Mind is One of a Kind!

Start Wising!

Generalizing -Analyzing- Internalizing -Strategizing - Actualizing!

Synthesis - Analysis -Synthesis!

Feel Your Belonging to the Digital Evolving!

2. Live Consciously and Visually. Be a Visionary!

Digital Intelligence is the present-day process of acquiring more life awareness and transforming yourself in a way which presupposes developing ***a new mental set-up*** and a new, transformed, and enriched intelligence.

It will enable us to synergize the negative and the positive sides in us to produce ***a holistically thinking human being*** *with much more advanced consciousness and limitless imagination. Albert Einstein* once said, "***When I examine myself and my method of thought, I come to the conclusion that the gift of fantasy has meant more to me than my talent for absorbing positive knowledge.***"

Everything around us is completely interlinked in time and space, and we need to get rid of our polarity consciousness, ***the illusory nature of duality***, generated by society-induced thinking. The modern times demand that we stop being channeled by *"the collective unconscious"* (*Carl Yung)* and develop our own **PIVITAL QUALITIES** in life - personal ***gravity, personal magnetism, and personal integrity***. Even personal power is dual. It starts with deciding on the one hand, and acting, on the other one, when problem-solving is separated in time and space. ***"Only with awareness of the whole, transformation and freedom come."****(Eckhart Tulle)*

Reprogramming the mind for the holistic, but individual perception of the world, we can empower the mind holistically, using ***both brain hemispheres of the brain in synch***, and expanding our vision both in space and time. New, **holistically developed intelligence** raises us to a higher level of consciousness that touches the ***Omega Point – True Enlightenment.***

The people who attain it can distinguish the terminal from the eternal and save our infinite souls from ignorance and impulsiveness of behavior. Knowledge empowers and refines with its limitedness. There are many great spiritual leaders now that we should follow knowingly, not blindly. I like *Drunvalo Melchizedek and Sadhguru* for their unique, well-processed, **OBJECTIVE KNOWLEDGE OF LIFE**. that reflects the vastness of our ignorance

Each of us represents a fundamental barrier to science attempts to understand and control the mind. Dreaming of finding ways of our expansion into space, imagining high speed travel in low pressure tubes, participating in the on-going robot revolution, or thinking of becoming a would-be specialist in robotics are the ways of expanding our digital intelligence.

Frankly***, digital intelligence also demands that we heighten our brain wave frequencies***! We are unable to see the aliens that are trying to contact our civilization because they are overtones higher in their vibrational power than us. The goal is to advance the operational capacity of a computer from billions to trillions, and from

trillions to quadrillions of operations a second. Consequently, we" *need to constantly turn the mental rheostat to the mark that is* ***"brighter, brighter, and faster!"*** *(Robert Stone)*

Mind should never Stop Steering the Brain, not just a Gadget in hand!

3. Brain Plasticity + Automation = Our New Self-Creation!

While scientists are working on programming computers with human capabilities, ***we must be self-coding new short-cuts that will be wiring our thoughts through space and time in a novel way.*** In fact, we can hardly imagine our life without different helpful gadgets at hand, the use of which adds up to the most amazing capacity of the brain - ***its plasticity***. *(Dr. Michael Merzenech).* ***Brain plasticity is the ability of the brain to adapt,*** to change, and to delegate different duties to other parts of the brain, should such necessity occur. **The systematization of these competences in five dimensions** is urgently needed as ***we are moving to the five-dimensional world...***

The ability of the brain to change our learning thanks to its plasticity is our unique capacity that will never be duplicated in a humanoid because its mechanism is God-set and will hardly ever be revealed by us. It is the Pandora box that will never be opened entirely. The AI creations only testify to **the unsurmountable power of the brain** and the necessity ***to develop it in a holistic way***, using the brain's amazing mechanism of neuroplasticity to create new connections in the brain and expand the parameters of artificial intelligence.

Computerization of knowledge goes along the way of creating quantum computers that will widen our intellectual horizons unpredictably.

Amazingly, the IBM researcher Dharmendra Mahathat and his team whose mind-advanced work we have mentioned above, have created a computer that is able **to route information *around a burned core in a computer,*** like a human brain does , delegating its functions to a healthy part in the brain. It is called neuroplasticity , and the wonders of ***Quantum Computing and Generative AI*** are possible thanks to this amazing quality of the human brain that now expands the capabilities of Artificial brain, too.

In his wonderful book" ***How to Create a Mind,"*** *Ray Kurzweil* is talking about "**the conceptual hierarchy**" of different activities that artificial intelligence is working on , programming the robots to recognize the patterns, stored in the memory.

The neocortex of the machine is trying to make sense of the input, presented to it. It is mind-boggling work, and with all due respect to it, we need to question the ability of artificial mind to operate the language with an amazing flexibility of thc thinking skills that a human mind displays on the path of its self-consciousness raising.

The Language Habits + Speech Skills that I am talking about in the **"Language Intelligence"** (*Section Two below*) leave a lot of other questions of the possibility of Artificial Intelligence to surpass us in this capacity. The most vital components of cognition here are the electronic and magnetic fields in the brain. They are generated by synaptic currents that constantly ripple through the brain and promote its evolution.

A Machine Mind is Monitoring its Functions like Brain Plasticity does, but not Better than Us!

4. A New, Humanized Mind Must Also Be Refined!

Artificial Intelligence has produced a new thinking mind that outshines us in many mental abilities, but one *- **our Language Intelligence and its evolution.*** *(See below*) I think that our **Language Intelligence will never be surpassed by any artificial intelligence** because the ability to operate the language is innate in us, and it relates to the evolution of our self-consciousness that is part of Universal Consciousness and "***that is operating our DNA.*** *"(Academician P.P Garyaev)*

A Part Cannot Surpass the whole!

All we should worry about is how to better program our super-computers with ***the language habits and speech skills*** that need ***to work in synch in a holistically wired brain and neurologically instilled soul.*** Humanoids are already becoming self-aware but not yet regulated in their innermost qualities. Our goal is to instill the best human qualities in them and stop the uncontrolled urge to destroy us. ***Naturally, robotic engineers must be holistically educated in sophisticated neural language operations.*** **Their cognitive skills must be integral with their self-consciousness** because the neural network that is instilled in them duplicates that of a designer or any sample brain that is far from being perfect in its moral ethical filling. **AIs might display self-awareness ,but they have no SELF-CONSCIOUSNESS that is our prerogative!**

The HOLISTIC FRAMEWORK OF OUR COMMON THINKING is a MUST HERE.

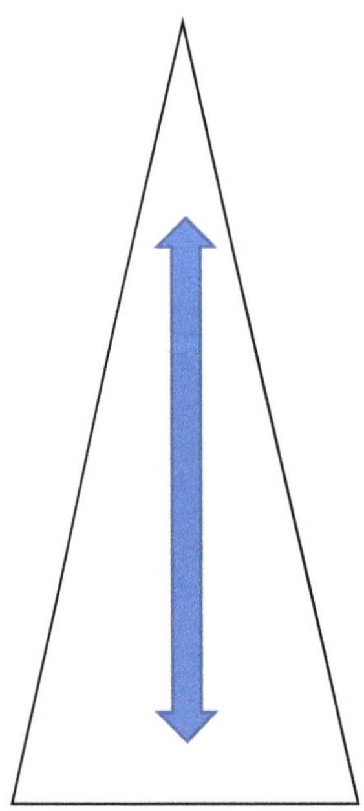

Super Consciousness
Spiritual Intelligence

Super-Conscious Mind

(***Concluding Synthesis***)

Intuition, telepathy in action, externalizing

Having intuitive insights

Generalizing *(inferring /concluding)*

Conscious Mind

Rationalizing / Reasoning/ Internalizing

(Analysis)

Inductive / Deductive *thinking)*

Conceptualizing / Creating

Analyzing / explaining

Subconscious Mind

Visualizing

(Initial Synthesis*)*

Seeing /Perceiving

Humanoids Mirror Our Own Innermost Imperfections!

"Robots are Taught to Be Better Robots. We need Humans to Be Taught to Become Better Humans."(*Jensen Huang, NVIDIA)*

5. Capitalize on Your Digitized Competences Consciously!

No doubt, digital evolution is great, and it is unthinkable to imagine our life without this leap in our intelligence. ***Digital intelligence gives us an evolutionary boost with the ever-expanding modes of communication.*** For the last couple of decades, people continually learn new things and ways to interact with each other on the Internet through English, called "***Globish***" now. Every day brings along new unbelievable changes that demand immediate adjustment of our electronic competences. **NOVELTY IS THE FEATURE OF NOW. WOW!**

Thanks to the evolution of electronics, each of us is building his / her own ***Tree of Knowledge*** which changes our belief system and impacts the way we rationalize our lives. Every concept, every opinion forms a little branch of that tree through "***intelligence networking***" *(LinkedIn)*The operating system of the brain is like a computer and, consequently, you need to take care of your brain, your "personal computer" against viruses and other pollutants ***to preserve your tree of intelligence intact and remain authentic in your thinking abilities.***

Digital intelligence has also come to the forefront of education, involving everyone, from toddlers to old people into the vortex of its interplay. Computers improve our cognition, sharpen multi-tasking abilities, speed decision-making, ***increase our aware attention span***, and even prolong our lives. The changes in electronics are amazing and often go beyond our ability to follow them because artificial intelligence is driven by *Thomas Edison's* challenge:

"If there is a better way, find it!"

This better way is connected not only with computation, but ***with the language competence*** that is the base of our thinking, especially in terms of holistically developing our brains. ***No doubt, digital intelligence strengthens our intellectual culture and develops our resourcefulness***

We are in a situation where we must devote much attention and time to deliberate ***the systemic development of our digital personalities, which is a new phenomenon in psychology,*** and we need to do it without self-imposed limitations and in synch with the demand of time.

Next, we see that ***physical intelligence and digital intelligence are inter-dependable,*** and their synchronic development explains why our indigo kids acquire these intelligences with amazing ease. Our children pick this new intelligence in the most natural way. Electronic gadgets help develop ***their holistic thinking skills through gaming intelligence visually and very integrally.***

Body + Spirit + Mind + Self-Consciousness + Universal Consciousness =

New connections in their brains are growing with an unbelievable speed, widening the generation gap that ***can be bridged because our language's thinking capabilities are much higher*** than those of our kids. So, let's keep growing in every respect in self-worth and self-respect!

When We Know Better, We Do Better!

6. Self-Awareness is a Wake-Up Call for Those who Want to Self-Install!

Every rhythm of life is the symphony to hear! Even the beat of our hearts has this rhythm. We need to consciously install the psychic protection of our inner symphony against evil injections of other people into our life with their whining, complaints, and negative predictions. Put your thumb at the pulse–beat on the left wrist and for a few seconds, focus on its balanced rhythm: **21-21-21**.

Calm down, calm down, calm down!

Humanoids will never get nervous, but they can calm us down in regulation our emotional set-up that might be instilled in them, they will work as our guardian angels, teaching us **EMOTIONAL INTELLIGENCE** .and instilling in ***us Emotional Diplomacy Skills.***

That is a great, evolutionary purpose of the thinking machines- modify our emotional reels.

The triumph of life in you will immediately calm you down and protect you with its magic. The AI can help you keep your **URV** (***the Ultimate Result Vision of yourself***) at the forefront of your mind as a screen. A clear picture of yourself as a totally accomplished person that you are becoming will do the trick. You will love yourself the way you are!

Self- Induction: **In my mind, I am One of a kind!**

Every page in this book ,concept by concept, is concluded with ***a rhyming and inspiring mind-set*** that is meant to focus your attention on the concept in question and serve as the short-cut to the brain at the completion of the page.

Thus, it gets reasoned out and memorized better. ***"A rhyming word gets better inward!"*** *(Edgar Cayce)* A robot -friend can be programmed to monitor your mental-emotional set-up with the most convincing, calming down, or mind-framing mind-sets and word them out now of your need. Any remark from a robot friend will never be perceived by us as a derogatory and down-grading one.

Electronically framed advice is better perceived by us. *The book provides an array of inspirational, psychologically loaded rhyming boosters to choose from*. You can pick them in accordance with your needs, mood, plans, and aspirations. Upload those that resonate with you most into your smartphone Use them if your mood sags and the problems of life dominate you.

Be physically, emotionally, mentally, spiritually, and universally upbeat!

Learn to boost your mood consciously and get back to reality, perceiving it calmly **AS IS,** and appreciating it, no matter what. Constantly boost your spirit with the inspirational mantra below.

Change Your Digitized Thinking, Speaking, Feeling, Acting, and Seeing! Be an Exceptional Being!

7. Immunity Against Digital Addiction

Obviously, computer intelligence that very many young, gifted minds possess, even when it is brilliant, is still ***mechanical, dead, programmed, and artificial.*** We should continue to expand our minds with the help of electronics, but we should not robotize them and become robots ourselves.

In the Universal Life Spine, we connect space and time with the Artificial Intelligence twine!

We must remember *that* ***the mind monitors the brain, not vice versa,*** and the computer mind cannot yet substitute for the human mind. It will be possible within two decades or earlier, and it will undoubtedly boost our mental evolution to new heights, but the problem is to find the way we will coordinate our self-consciousness development with that of humanoids.

Consciousness is infinite while the digital mind is finite, created by Infinity in action. The downside of our overwhelming enrolment with digital evolution is our inability to prioritize the essential intelligence and develop them holistically, using the electronic intelligence as a backup tool for developing our self-consciousness! That's what it is here for!!!

Research warns us that using social media, ***we get used to it in terms of drug-like behavior***. We get overwhelmed with the avalanche of information, and we often become life-smitten with its flooding stream. We become worn out, de-focused and very virtually entangled emotionally. Consequently, we fall into ***the state of commonly inflated, or deflated thinking*** when we are unable to think straight and be consciously detached from computer programming.

Be Cautious! Be mindful; Be conscious! Feel but think!

I repeat again that the process of developing true and aware digital intelligence must have a holistic outcome that works for the development of our higher consciousness. It should not be hampered by **robot-like thinking** that becomes totally dependent on the machine because it shoots into the brain **DIGITALLY STAGNANT** habits. I witness my students' inability to ***get though the wall of blurry thinking*** that becomes a new educational challenge.

If the flood of digital information is not properly digested, it clogs the mind. *Information Processing Skills* ***must be developed without any delay.***

Undigested intelligence results in situations when a student does not assimilate knowledge; He /she cannot ***disassociate himself / herself from virtual reality***, cannot feel mind-free to express an individual opinion, not the one that can be easily plagiarized and perceived as one's own

The thinking skills enhanced digitally, need to be processed through the holistic paradigm of – ***synthesis-analysis -synthesis*** *(See above*

To keep your consciousness intact, practice Auto Suggestive Meditation. consistently.

Use Technology for Your Self-Ecology!

8. Digital Clicking in Love-Seeking

Automatic thinking without awareness is the main dilemma of human existence now! Unprocessed digital intelligence deflates a mind-heart connection if such clicking ever occurs! Numerous match services are booming, but heart + mind connection is swooning. Everyone is hoping for a soul mate, being unaware that clicking should occur at five holistic levels, or at least in three - ***physical, emotional, and mental,*** preferably starting with the spiritual probing of each other and finishing with the physical testing. ***Change the frame of thinking to a conscious one.***

Doing the right thing polishes the mind - heart ring!

The framework of many sights for dating is set in scanning a potential partner by asking to fill in the questionnaires that are psychologically unprofessional and superficial. ***They need to be holistically targeted in five dimensions of life, mentioned above.***

Digital engineering offers girl -humanoids for service, and it is, of course an amazing improvement of the sex toys. But such creations impact **heart -to-heart** and **mind-to-mind** interaction between men and women that reinforce their automatic, impulsive habits and ***destroy the authenticity of love completely***. Other than that, such sex and pornography , like fast food, become too addictive. ***They generate more mind-heart disconnection that we are supposed to repair.*** So, think before you get addicted to electronic sex partners.

Too specified questions are not helpful here. The quality of our intelligence must be in an inseparable connection with the ***heart-based qualities*** of kindness, compassion, care, and consideration for the other need to be considered. **We should click on both planes to have love gains.** Thinking + feeling awareness is the blueprint of self– creation and self- fairness.

Being aware means getting rid of compulsive thinking and living in the past. Since our relationship chances are limited to digital communication increasingly, we need to use digital means to ***enhance, not deflate the mind + heart connection.*** . *Very soon robots will be able to duplicate and even surpass us in our unique ability to love. The key* to this strategy is our understanding of the power of the opposites and the ability to connect them in our behavior, *synergizing our joint intelligence and consciousness on the way.*

Self-change focused on emotionalizing the heart and intellectualizing the mind is the solution in every area of our technologically enhanced life.

In sum, to rationalize and humanize our lives, we need to learn the cause-effect strategy of living and enhance the mind + heart connection. **Transformation of consciousness to a higher level must be the goal of our inventive minds in technology**. It presupposes the change of the frame of thinking, on the one hand, and our observation of ***the cosmic Law of Cause –Effect***, on the other. Philosophy teaches us to constantly clean the dashboard of our life errors and start from a clean slate. Our kids should be taught to observe this Law at every level of intelligence formation.

"Eliminate the Cause to Change the Consequences." *(Richard. Wetherill / " Right is Might")*

9. Raise Self-Discipline on the Digital Scene!

Digital intelligence is the greatest opportunity for us to establish the needed connection between **THE MIND AND THE HEART in GAMING** business, which is an open sky for us in this respect. .At present, electronic games is the field of warmongering and the display of super-human powers that develop wrong addictions, while the gaming opportunities are sky- bound in terms of helping to develop **AWARE ATTENTION** to professional intelligence*, or* the kid's / teenagers' / grown-ups' abilities to manage themselves.

Equally important is to use **Artificial Intelligence** + **Quantum Computing** to cultivate good human qualities and lay the foundation for a moral, ethical framework of the innovations in neuroscience and digital technology of the future. We need to instruct our kids and ourselves to love, care, display compassion, kindness, truthfulness, and love, through games that often heighten their aggressiveness , careless ,and an impersonal attitude to the outside world. *Such games only generate inner imperfections* and disconnection with the parents because the participants identify themselves with the electronic personages and are losing their own individual characteristics.

Virtual reality gets extended into real life that is perceived as hostile by the brain that needs re-programming **physically**+ **emotionally**+ **mentally**+ **spiritually**+ **universally** **in a tandem with Generative AI**

Body + Spirit + Mind + Self-Consciousness + Universal Consciousness =

Psychologists are desperately seeking the solution to the new addictive behaviors of the growing souls that, when they reach their teens become unmanageable, disrespectful, and impersonal, seeking more fun in drugs, alcohol, and excessive, uncontrolled sex. (***Holistically applied awareness at*** every level generates a new quality of **PERSONAL INTEGRITY** that keeps a person from getting morally de-magnetized.

5. ***Universal Awareness*** *999 (+)*

4. Spiritual Awareness

3. ***Mental Awareness***

2. ***Emotional Awareness***

1.Physical Awareness (-) 666

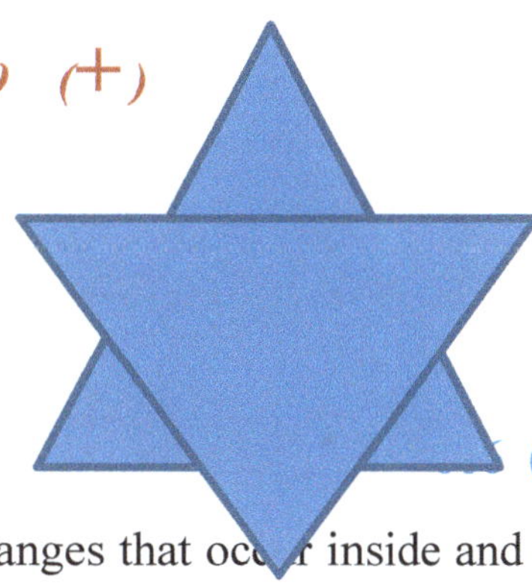

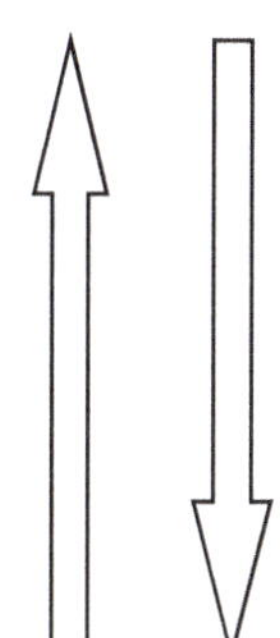

We just need to be conscious of the changes that occur inside and outside us, getting much more aware of these changes and much more informed about the reasons that generate any emotional or psychological consequences for us.

The Cause-Effect Law must rule the Mind's Store!

Humanize and Harmonize Your Contaminated Inner Device!

10. Be Integral, Sincere and Kind! Be One of a Kind!

In sum, **DIGITAL SELF-MONITORING** is a fantastic opportunity for us on the path of raising self-consciousness and an immense help to our psychologists. Future avatars will be attuned to our ***psychological imbalances***, giving us proper warnings, and helping us get together at the right time

Start visualizing your spiritual, psychological, mental, emotional, and physical blocks. De-magnetize them and delete them with a conscious effort consistently. Be *Medusa Gorgon!* *Freeze stone* ***the unwanted patterns of behavior, emotions, words in you!***

Psychological awareness (*See Psychological Intelligence below*) is essential in electronic games *to* widen the scope of the beneficial effect of the games that are often the war-fields of aggressiveness and the demonstration of the super-power. There are boundless opportunities to engage players in the activities that develop ***Holistic Living Intelligence Skills*** and instill the most noble qualities in them. Being integral, sincere, and kind should reside in a superhuman mind!

SACREDNESS +NOBLENESS + LOVE must become HUMAN+ AI ethical stuff!

In sum, we should use our future electronic friends, programmed into a wristwatch, or any other mind-boggling form ***to boost the spirit*** by suggesting right decisions, expand our educational horizons , or instill in us the right emotional response by charging the situation-prompted mind-set at the right place and at the right time.

For sure, future gadgets will be finely tuned to our mental, emotional, and psychological states. Electronics is our step forward on the ladder of evolution, and it is the greatest device that we have had so far in our exploration of the world. Shape your digitally focused mind to become more self-consciously refined!

Marvel at the Internet that is God-set but be connected to the Conscious Life Net!

Auto-suggestive mind-programming does just that. You can program your smartphone not only to boost your spirit, but also to instill in you the qualities that you wish to enhance at any time and in any place. Very soon, we will be able to fix our psychological and moral imperfections with the help of digital identities, called ***Avatars***.

They will reveal more positive aspects of our personalities, making us more self-controlled and goal oriented. Your auto-induction is:

I Can…! I want to…! And I Will…!"

Undoubtedly, ***digital intelligence will promote our brain wavelength to a much higher overtone,*** enabling us to heighten the level of our self-consciousness with each evolutionary step of humanity at large.

Wow! I Wish I Could Live then in the Unanswerable When!

(End of Stage One- Physical Intelligence Part)

Stage Two – Self-Awareness

Mini-Level of Intelligence

Language

Intelligence

"There is only one language in the world – the language of God –
The rest are just dialects"

(Umberto Eco)

(For more, see the book "Language Intelligence or Universal English")

(Www. language-fitness.com)

Be Language-Fit to Feel Complete!

Have Language Awareness with Fairness!

Design by Yolanta Lensky/ my daughter

Language is Me; Language is My Identity!

The Intellectual Odyssey of a Seeking Soul

Vistas of Intelligence to instill next:

Inflating Your Personal Pride with a New Language Might!

10. Universal Intelligence — ***Self-Salvation***

9. Spiritual Intelligence — *(Universal Realm)*

8. Social Intelligence — ***Self-Realization***

7. Cultural Intelligence — ***(Spiritual realm)***

6.Financial Intelligence — ***Self-Installation***

5. Professional /Creative/Intel. — *(Mental Realm)*

4 Psychological Intelligence — ***Self-Monitoring***

3.Emotional Intelligence — *(Emotional Realm)*

2. Language Intelligence — ***Self-Awareness***

1.General Intelligence — *(Physical Realm)*

/ Physical / Cognitive / Digital /

Language is the Portal of our Common with AI Thinking!

To Make Your Life More Digitally Wise,

Generalize +- Internalize + Personalize + Strategize+–Actualize!

Be Wise

Don't Be Language Negligent. Be Language Intelligent!

1. Language is the Matrix of Human +AI Thinking!

Every Language is a miniature universe and language intelligence determines our life in it. Each language phenomenon is rooted in the brain, as the matrix of thinking and the root of our self-consciousness. In fact, *"**pure reason**" (Emanuel Kant)* ***is the Universal Intelligence*** by which every language phenomenon is roofed. Therefore, o***ur language capacity development will always exceed that of a humanoid in the quality of its thinking and its conceptual display***.

Language Intelligence separates us from the animal kingdom, and it gives us the main assert – our THINKING CONSCIOUSNESS. ***"Thinking is a subtle state of speech, verbal or written."*** *("Language Intelligence or Universal English," Book One)* Rationalization of the language use is inseparable with the rationalization of thinking that requires another level of **LANGUAGE AWARENESS** and language correctness that inevitable affect our thinking.

It's very important to remember that being the global language, English is developing in people, learning it the **SECOND LANGUAGE PERSONLITY of** a very special, uniting character because it teaches the people to learn it for the global communication not only to think in a new, digitally enhance way, but to also **mold their personalities**, adjusting them to the new exponentially developing digital reality, less impulsively and more consciously.. **Life is tough, but I am tougher!**

The words we use create the state we are in. Inspiring words inspire! *Despairing words despair!* The most vital thing is to think for himself / herself. That's a big mistake of many parents that stereotype the thinking of their kids. ***It is a mentally criminal act*** that is being invested by school education, mass media, and society overall. There is no intelligence of any kind without language in the brain!

That is why language Intelligence is ***at the mini level of the holistic pyramid of intelligence.*** I have not started the presentation of **the hierarchy of intelligence** with language intelligence because, unfortunately, before a person comes to the realization of the necessity ***to tame his tongue***, he first gets interested in preserving his health and prolonging his life span. No doubt, language Intelligence permeates all the vistas of intelligence.

Language is the mind; they are an inseparable celestial index!

Above, I have compared ***the interconnectedness*** of all the intelligences that we must develop on the path of self-formation to ***the Russian dolls.*** Like *the Top Doll* that encloses all the rest of the dolls in its structure, Universal Intelligence is governing our intellectual development and consciousness rising, acting as ***the blanket intelligence*****,** with ***Language Intelligence*** at its core.

Language Intelligence needs to be developed holistically, too, because it is ruling our **COMMON MIND** now**.** English, as a global language, is unifying us mentally, forming **UNIVERSAL SUPER-INTELLIGENCE** in all of us with ais help.

Language is the Core of our Cerebrally Conscious Existence.

2. Language Awareness without Negligence

Our consciousness is part of Universal Intelligence / the consciousness that will never be exceeded by any machine because ***no machine will ever be able to operate the language*** the way our trained brains do. ***Language Consciousness is at the start of our human evolution!***

"Language is a dynamic interplay of consciousness." *(Rav. P.S. Berg)*

Developing our intelligence, we are automatically raising self-consciousness, and both evolutionary accomplishments are impossible without the holistic development of language intelligence, or the unity of ***the language habits and - speech skills*** that determine holistic thinking and can be improved with technological enhancement. Holistic thinking, which is the work of both brain hemispheres, can be instilled in our learners only based on holistically developed language skills. The brain is an overly complex machine, but its broad topography is simple and symmetrical in managing the two main parts of it. The left hemisphere is *rational, analytical, and logical*; the right one is **instinctive, intuitive, and systemic**. Language resides in the left hemisphere that reasons sequentially ***excels at analysis***, oversees words, and ***operates the structural system of the language, or its grammar.***

The right side of the brain reasons holistically, recognizes patterns, ***interprets emotions, and builds up thoughts and concepts with the vocabulary tools*** that the left hemisphere provides. **That's why the development of this hemisphere is crucial in technological times**. The right hemisphere of the brain is underused at present, and this fact undermines our holistic brain development. Obviously, knowledge in neuroscience in this respect is essential for boosting our thinking abilities because driving a car blindly, we will eventually come to a dead end. ***Language awareness boosts language intelligence***.

Language Intelligence develops Language Consciousness!

Language Consciousness is consciousness itself because language frames our thinking, and it is the pe-requisite for cognitive development of the brain. A person who hardly ever reads and is tongue-tied cannot accomplish a lot in life on the venue of creation. ***So, consciously developing our language intelligence, we are automatically raising self-consciousness that is the goal of our evolution.*** It is a great accomplishment of digital technology, its greatest miracle!

Language is mechanically programmed now, but it should not be mechanically dynamic in raising our self-consciousness that is supposed to evolve the **holistic mind + heart connection** that is broken in us at present. Evolution is pushing us to this unity, and we can witness it in the best, most advanced people among us. This unity is the reflection of the holistic development of the brain with the right hemisphere taking the creative lead. We cannot change the outer projection ***(speech)-*** **the right brain** ***(language)-*** **the left brain.**

Being Language Conscious, we become more Life and God -Conscious!

The Whole Brain Intelligence is Our Evolutionary Goal.

3. Language Consciousness Forms Our Intelligence!

Unfortunately, **our language intelligence is suffering tremendously now.** The memory bank is clogged with disunited information; the vocabulary stock of many people is appallingly limited. There are many junk words and phrases that ***substitute the clear-cut expression of thought. There are no words in the brain to help a primitive mind out.*** The hierarchic nature of language mirrors the hierarchical nature of our consciousness, or unconsciousness.

Language Intelligence is Language Consciousness in action!

Our language consciousness is part of our ***Universal Language Interaction*** with ***the Unified Field*** of ***Information*** *(Dr .John Hagelin)* that is permeating everything that we are part of. The scope of our memory can retain seven or nine pieces of information or words at one presentation. **AI + Quantum Computing** ***enhancement will change this capacity tremendously***, but the brain won't be able to ever work at the speed of the hyper-computers that we develop. It is quite natural and understandable. Quantum Revolution is our Language Consciousness Solution!

_ Since artificial intelligence disciplines the use of the language, ***we need to do the same***, but in our creative, conscious, and exponentially unstoppable way because we have ***flexible, not mechanically based language-intelligence***. Our interaction is based on language -consciousness, inseparable with the Unified Field o Super-Consciousness, perceiving our mental signals and responding to them through its digital means.

Language intelligence is" telling" the brain not only WHAT to think, but HOW to think, too.

As a psycholinguist, I am sure that computers will never be able to beat us in our language intelligence and the flexibility of ***thinking*** + ***feeling***. capabilities. Computers can create art, compose beautiful music, inspire great minds of *Steve Jobs* type, but they will never produce another *Lord Byron or Tchaikovsky*, another *Einstein or Dostoevsky*, another Sigmund *Freud or Nikola Tesla.* T***he consciousness of a genius is inseparable with the unique language intelligence***, inseparable with universal Super-Consciousness.

Body + Spirit + Mind + Self-Consciousness + Universal Consciousness=A Whole You!

Such people had exceptional ***minds and an extraordinarily rich emotional perception*** of the beauty of the world, that the artificial mind will never be possible to develop in the **HEART+ MIND** sync. Most importantly, artificial intelligence can never master cognitive and emotional intelligence holistically which is another step-in **LANGUAGE CONSCIOUSNESS**,

The mind-set that is below is impossible to be instilled in AI installed machine being!

Intellectualize your Emotions and Emotionalize the Mind. Be One of a Kind!

4. Language Fitness without Cognitive Sickness!

We know that *the **digital neocortex is being created in us now***, but I think that, irrespective of this mind-boggling fact, we won't be able to raise our language consciousness that I have touched upon above , without healing our ***Language Habits*** and ***perfecting the Speech Skills,*** both in a native language and English that is called by linguists "***Globish***" now. The way the English language is used on the Internet and while people are texting on their smart phones is appalling.

Language fitness is directly connected to RATION + EMOTIO fitness!

The digitally enhanced evolution demands **LANGUAGE FITNESS** from us now**. LANGUAGE LITERACY** is violated, and language negligence is killing our holistic thinking development. ***New language awareness needs to be instilled in our schools as early as possible***, because the damage that we are doing to our language intelligence might become irreversible.

We must be language-fit to succeed! *(See www. Language-fitness.com)*

The English language, when consciously used, can express absolutely anything in the most precise, thought-provoking, insightfully beautiful way because it has an exceptionally simple, linear grammatical structure. English is the first global language in computation because its structure is a one-way street.

English is a mathematically perfect language and, therefore, it can be used for programming. The word-order of the English language lets the machine perform linear thinking operations. The lexical wealth of English has an incredible volume of a shot-cut scientific preciseness, too. The scientific articles written in English are the most precise in their conceptual structure. That is why exact translation is never possible computer-wise from English into any other language and vice versa.

The linear restructuring of the mind is needed when the English mentality comes into play. That is why English is rapidly developing from the status of a global language to the status of the universal one, ***the language of God!*** Obviously, language operating robots should be programed by incredibly good linguists, not just electronically smart programmers. We should not have any negative experiences language-wise, like with poorly programmed dictionaries.

In sum, we need to raise our awareness about how the brain works , creating the mind. ***"Our neocortex is virgin territory when the language mentality is created. It has the capacity of learning and therefore creating new connections between its patterns, but it gains these connections from experience"*** *(Ray Kurzweil) It means that learning and recognition take place simultaneously. If we learned the pattern consciously, we recognize it. The neocortex is continuously trying to make sense of the input presented to it.*

So, we need to constantly select the information we get and delete the one that clogs the neurological paths and blocks our creative thinking and holistic brain development.
Consciously discipline your thoughts language-wise!

Language is the Passport of Your Personality!

5. Language Pollution Impedes Personal Evolution!

The lack of language controlling skills is always connected with ***the lack of bad language immunity*** that I have touched upon above. In the process of self-transformation, we cannot afford to speak casually, automatically, polluting our speech with banal phrases, junk words, and commonly used misjudgments.

We must not let ***our bad language habits make us dis-empowered emotionally and mentally***. We must kill them because, as Albert Einstein put it,

"Bad habits have a good tendency; either you kill them, or they kill you."

Good knowledge of a native language or of English is marked by our thinking, feeling, and acting in these languages. There is a flood of **LANGUAGE POLLUTION** on the Internet and around you, but our language intelligence should not be affected by ***the avalanche of contaminated language***. There should not be any casual word use, without any monitoring of a thought. A beautiful British saying testifies to such a necessity.

Self-induction ***"You never have a second chance to make a good first impression!"***

The holistic approach to language-speech performance also demands that we develop tight control over **the synchronicity** between the language habits and speech skills that we have noted above. It's not an easy thing to master, but it is not altogether impossible.

LANGUAGE HABITS (***Pronunciation, Grammar, and Vocabulary***) **that constitute the language operandi system** are formed very slowly, and they require aware attention to be paid to redesigning them to make our pronunciation or diction comprehensible, our grammar accurate, and the vocabulary use varied and exact in its transmitting our thoughts.

SPEECH SKILLS ***(Listening, Speaking, Reading, and Writing)*** are formed much quicker. We need to learn to be economical and exact in speaking, conceptually oriented in reading, and insightfully perceptive in writing. However, if the speech skills are formed on the automatic pilot, without **aware attention being paid to the process of conscious language use**, unhealthy habits are difficult to eradicate.

That's why many junk words that pollute the speaking of our kids are so difficult to be rid of. Bad language rituals result in bad speech performance. Bad speech performance results in a bad destiny because we stop paying conscious attention to the way we speak and to the exactness of the messages that we communicate with the help of the language.

Bad language control and meaningless talkativeness generate family problems, relationship fractions, job problems, money loss, and health worsening. Only conscious, aware attention to the way you speak does the trick.

Bad Language Habits Result in Bad Life Rituals!

6. Correct Language Framing

Since speaking is our first action since birth, it needs to be governed by this rule all life. So, don't ever stop controlling your language, a native one or English, because the image that might have been once sharp and clear becomes dim and blurry with lack of language control. What we do not use, we lose!

A Minute of Thinking is Worth an Hour of Talking!

Language degradation is a very quick process that drags us down unstoppably! ***Degradation in language intelligence is also connected with the degradation of a personality*** - slagging of character, weakening of will-power, emotional misbalance, and **INFLATION OF THINKING.** *(Cognitive Intelligence, Chunk 5)*

So, motivate yourself to do any action by correctly speaking about it. Any problem gets resolved through **CORRECT LNGUAGE FRAMING** because any problem awareness is heightened by language exactness and the emotional control of it.

A soft answer turns away the disaster!

The philosophical effect of framing is amazingly effective in preserving our language ecology and dealing with language intelligence, emotional intelligence, or psychological one. An unframed picture always looks less impressive. Professional language has a special aura, and it is , of course , justifiable because it serves as a short-cut to the brain is the professional problems solution.

However, we have polluted our most needed to be developed **HEART + MIND**. Link with the *bureaucratically deaf language* that needs a special interpretation to be employed.

Auto-induction: **I am much less if I am LANGUAGE- FRAMELESS!**

In short, language pollution impedes personal evolution. What's amazing is the inner feeling of accomplishment that you will generate once you start working on your **language ecology consciously.** ***Self-Synthesis - Self-Analysis-Self-Synthesis is your language paradigm, too. It is your language-based thinking that helps your wholeness inkling***! (Www.langauge-fitness.com)

Body + Spirit + Mind + Self-Consciousness + Universal Consciousness

Be devoid of automatic speaking negligence. Respect the language of Universal Intelligence!

The orchard will bloom in a blast if you say to yourself fast:

I CAN…! I WANT to…! and I MUST…!"

Auto-induction:

I Commit to Being Language-Fit!

7. Language Correctness and Speech Exactness

"Language is our first action in life!" *(A. A. Leontief)* There is no evolution of intelligence without ***language fitness*** because there is no development of consciousness without the clarity of thought and the accuracy of words. Also, language has a profound effect on our life, and it needs to be grammatically correct and lexically precise. We are defined by the way we speak and write!

Language should be in rapport with speech in the holistic unity of the left +right hemispheres.

The moment a person opens his mouth and says something, you can place him for his social status, profession, attitude to life, and even his philosophy of life. Taking care of our physical fitness, or our physical intelligence, we often forget to control ***our verbal impeccability*** and, thus, we display the most disrespectful attitude to the language in which we communicate our thoughts.

In our digital times, the adequateness of the thoughts that we impart and the language that we use have become appalling. We cannot establish **a good mind-to-mind contact** with a conversation partner in face-to face communication or in virtual space, unless the language and speech are in synch.

We need to speak / write correctly, language –wise, and with precision, speech-wise.

The correct speaking of a person relates to his correct behavior because it signifies his correct thinking. Speaking with precision and, therefore, correctly, being able to adequately communicate your ideas requires **a precise choice of words** that gives you the chance ***to say what you mean and to mean what you say.***

Language unfitness always points to lack of intelligence.

We should note here that any personal contact to be adequate must be based on proper language habits – **good grammar, correct sentence structure, and a meaningful vocabulary choice.** We must sharpen our language skills all our life to be able to express ourselves clearly and consciously, both verbally and in writing.

The use of the language in artificial programming should be especially accurate. Language operating should be marked by conscious, controlled, sense-making mode of speaking that requires much richer vocabulary stock, too.

The brain constructs the vocabulary of a single person.

It is always personal, and therefore, each of us speaks and thinks differently, and this difference is always monitored by the richness / scarcity of the vocabulary that we use. It will not be an overstatement to say that there is no personal growth without a substantial growth of the vocabulary stock in a person's brain.(

See *"**Language Intelligence or Universal English!**"2021)*

The Richer the Vocabulary, the Better the Thinking!

8. Practice Visual Listening / Speaking!

We can diversify our language habits and speech skills much better if **we read consciously**, paying attention to what we read, ***conceptualizing the text while reading*** and paying aware attention to the richness of the vocabulary that is being used by the writer. These skills should be developed in schools through extensive ***mind-provoking reading though, analyzing, interpreting, and synthesizing the information*** *verbally and in a written form.*

Synthesis- Analysis –Synthesis!

Professional vocabulary skills are also extremely important in this respect *(See Professional Intelligence)* The professional use of the language means that we need to make the right vocabulary choice to express our thoughts adequately in any professional environment and to describe any problem in a concise way and to find a shortcut to its solution as quickly as possible.

Self- Induction: **"Talk the Talk!"**

This rule is applied to any area of expertise because every profession has its own vocabulary stock in use. If we are not aware of this vocabulary, our talking about job problems is vague and, hence, the decision-making skills will never be formed adequately. It is also extremely helpful professionally ***to visualize professional terminology*** when you hear it not to fall out of the loop.

The rule " **If you don't have something nice to say, don't say anything at all!"** is extremely important in a multi-cultural world with diverse personal characteristics, religious beliefs, and values.

Our collaboration with AI will help us follow suit and remain respectful in any situation. Evolution of language use requires ***the paradigm shift in language competence*** because it redirects our attention from the predominant focusing on speech skills to a holistic and keenly conscious linking of ***language habits and speech skills.*** *(Chunk 2 above)*

To Life-Succeed, Be Language-Fit!

We can tell our brain not only what to think, but how to think and what to say! Going along this path of **perfecting our language intelligence**, you will inevitably stop being self-conscious, especially when speaking in public and become more language conscious.

With your attention to the language in you, you will be much better perceived by people in your face-to-face, mind-to-mind, and heart-to-heart communication. *(See Language Intelligence or Universal English, Book 3)* ***Communication on a heart-to-heart and mind-to-mind basis is an opportunity to be changed by another person who speaks a different language.***

Stop Seeing Yourself in the Language and Start Seeing the Language in You!

9. Language-Speech Immunity

To conclude this part, language intelligence also means developing the **LANGUAGE-SPEECH IMMUNITY** against using harsh words, junk words, cursing, insulting people, yelling, constant lying and, most importantly, judging yourself and others. The ability to control thinking and speaking is a real might! Bad language-speech immunity is an obstacle to self-regeneration overall. Language pollution is polluting the mind, and therefore, a person's self-consciousness.

Auto-induction: **Watch your thoughts to watch the words!**

Bad language habits get harbored in the brain, and they require **individual**, **intentional, and conscious surgery** to be done to irradiate them from there. In his wonderful book that is amplifying to us the meaning of the Universal Law of the Right Human Behavior, Richard W. Wetherill writes," ***"Nobody can change human nature, but we can change people's thinking and speaking and though that their behavior can be changed, too."*** (*"Right is Might"*)

Language de-contamination and its eventual healing is a willful process!

Try to have a small language survey of people speaking around you. Go for a walk in the park or just sit on a bench and watch the world go by, focusing especially on how people are talking behind your back or in front of you when passing. You will be appalled to discover that they are talking about someone doing something wrong, or they are complaining about something. Negative assessments of people and life are heard all over: ***"That's absurd. / It's insane. / It's crazy. / It's ridiculous. / I can't take it anymore! / Are you nuts? / Leave me alone. / You know what I mean, etc.*** There is a lot of cursing accompanying these reactions, too. **No Soul-Symmetry!**

Body + Spirit + Mind + Self-Consciousness + Universal Consciousness =

Why can't we share our most inspiring emotions, talk about some wonderful things that we have done, complement each other and oneself on doing a lot of good things that often go unnoticed and unappreciated because we are not in the habit of paying attention to them?

I am extremely grateful to ***Oprah Winfrey***, **the woman that had shaped the minds of millions** in the world and injected eternal, positive values of love, respect, and morals into her public

It was the first TV show that I was very much impressed with, being very used to the Russian style of intellectualizing the public through TV. People hunger for the shows of this level of ***intellectual challenge and emotional pumping.***

What a great professional and personal magnetism and inner integrity!

Obviously, the quality of thought generates the quality of speech and modifies behavior.

Intelligence and Personal Altitude are shining through Your Language Attitude!

10.Tame Your Tongue Not to Be Mind-Stung!

"Language is impregnated with individual characteristics of a person" *(John Binge)* People who are language–savvy get along better with relatives, friends, and colleagues at work. Their thinking is normally well-structured, and they can better communicate their thoughts to their listeners. They think before saying something, and they oversee their tongues.

A **language-savvy person has a strong, confident, well-developed language personality**.

Goal # 1 – **Develop a new language personality- the primary goal in working on language intelligence.** A language-intelligent person manages his language as much as he manages himself. Every religious book of wisdom tells us about the absolute necessity ***to tame the tongue*** since uncontrolled speaking is sinful. In our knowledge–based and increasingly competition–driven society, ***language skills are the number one aptitude*** and determines our intellectual health.

Auto-induction: ***"Be impeccable with your word!*** *(Don Miguel Ruiz)*

Goal # 2 - **Language Intelligence means thinking first and saying something next.** If you become polished in your language in tone and vocabulary, if you are very outspoken and clear-cut in self-expression, if you can ***express yourself clearly the first time around***, your language intelligence is healthy and solid. All these qualities become attainable if you manage your language, or, in other words, *if **you put the letter" T" before the letter "S"*** in your language use.

Auto-induction: ***Think before speaking!***

Goal # 3 - **Redundancy pollutes our communication and kills our precious time. *Speak less*!** Language mirrors your personality, and it helps establish and maintain a better rapport with people if you are clear, concise, and exact in self-expression. Soul-Symmetry preservation must be in mind all the time!

Self-induction: ***Avoid speaking with nothing to say!***

Goal # 4 - **Learn to hold your tongue to avoid saying more than you can!** Language communication needs an absolute establishment of a **mind-to-mind contact** with an interlocutor, especially now, at the time of digital communication. Only conscious and concise communication teaches a person *to recognize **the relationship within the structure of thought.*** Only then can a person end up saying what he / she means to be understood adequately.

Auto- induction: ***Don't verbalize everything you want to say!***

Goal # 5 – **Say less but mean more! Learn to decode implications!** *To* achieve this goal, we must change our cognitive culture through rationalizing our thinking and learning to structure it properly. This is the language skill that we desperately need for our new electronically based and choice-driven economy. Speaking and writing with implications is a great gift!

"The method of the tip of an iceberg" was introduced in the world literature by a great American writer Ernest Hemmingway. The tip of the message, like the tip of an iceberg that is seen over the water, is said, while the rest of the iceberg or the body of the message that you want to communicate is under the water. ***It is implied, and it needs decoding.*** Such communication is very economical, but it necessitates a great deal of awareness of the subject of the conversation. This is what is required in the professional intelligence development nowadays. (*See Professional Intelligence, Stage Five)*

Goal # 6 – Language respect is also Self-Respect! It has become a bad habit for many people to assess things negatively and to complain about being discontented with the state of things in their life, or the life of other people. ***Judgmental speaking*** has become a norm among young people, too. Every other word in their conversation is "*,**She is like…, he is like….***" behind that person's back. Listen to the people talking around. They try to impress their listeners with something out of line, judging and condemning someone or something, expecting a listener to back their judgment up with the same train of thought.

Goal # 7 - "Clever people talk about ideas, mediocre people about events, stupid people about people."(*Rubaiyat of Omar Khayyam)* Judgmental speaking is being instilled in us by mass media that focusses only on the negative side of life for its sensational value.

Why don't we learn to notice the positive, constructive things first, and respect any constructive gesture, especially on the part of the President of the Country whose job, being the hardest of all, is judged most of all.

Also, **the uncontrolled language behavior** of people in some TV shows and their manners are accepted by the public as a norm. The language of the participants of such shows is beyond description and ***the inflation of their*** *language* ***skills*** signifies the inflation of their thinking that's hardly curable. Nevertheless, such shows are appallingly popular

For more on language fitness, see www.language-fitness.com and the books " Language Intelligence or Universal English

" **(Book One,** *(the Method of the Correct Language Behavior***) Book Two** *(Remedy Your Language Habits)* **Book Three** *(Remedy Your Speech Skills)*

*Also, see the book "**Americanize Your Language, Emotionalize Your Speech!"***

You and the Language in You are the Inseparable two!

At Universal Bay,

You are What You Think and What You Say!

End of Stage Two - Language Intelligence

Stage Three

M eta-Level of Intelligence - ***Emotional Dimension*** – ***Self-Monitoring***

Emotional

Intelligence

Inflate Your Personal Pride with Emotional Might!

Auto-Induction:

I am Not Emotionally Negligent. I am Emotionally Intelligent!

Change Your Internal Ritual to an Aristocratic One!

Emotional Intelligence with Much Negligence!

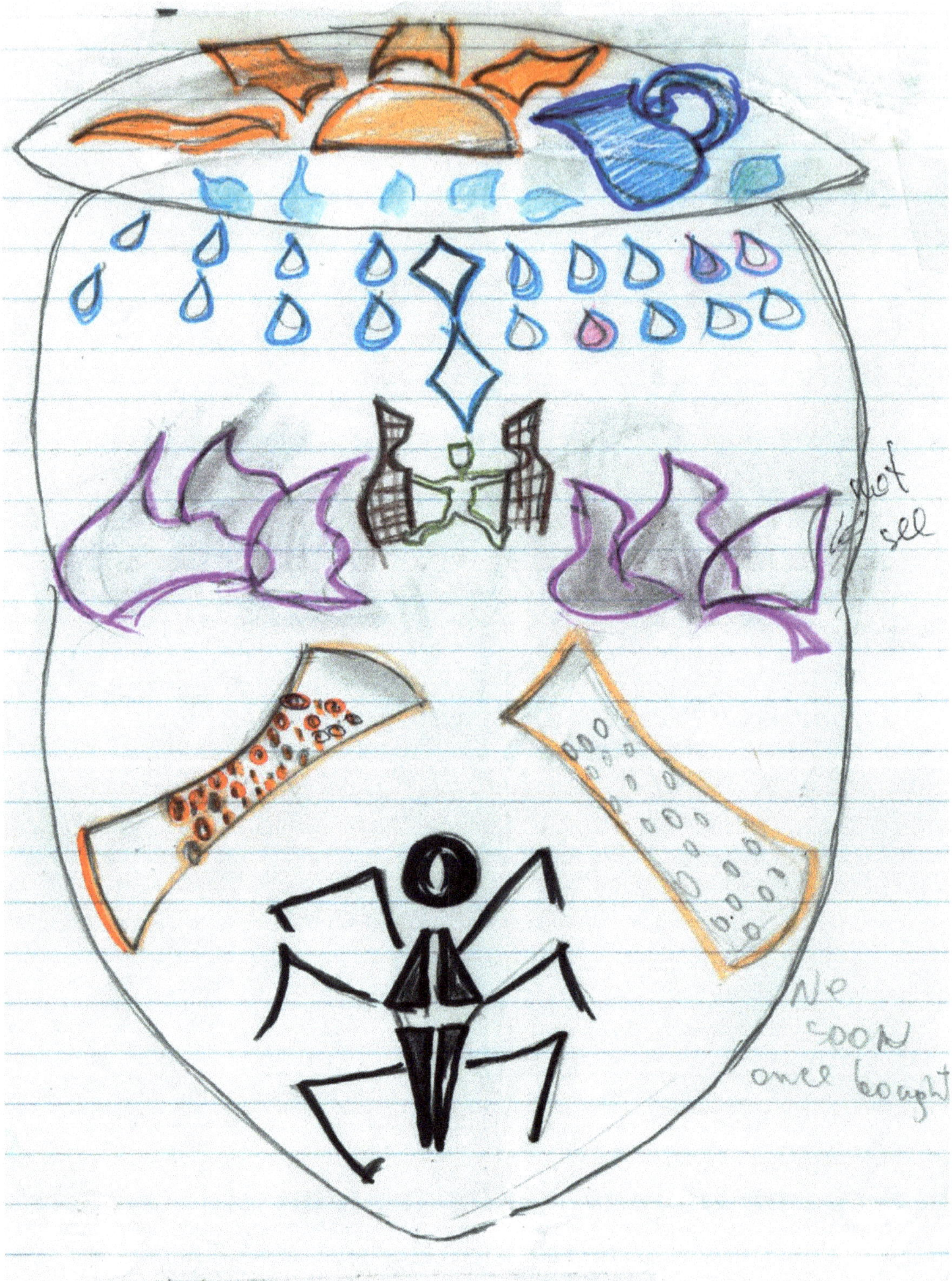

Harmonize Your Inner Buts to Enjoy Life's Guts!

The Intellectual Odyssey of a Seeking Soul

Vistas of Intelligence to install Next:

10. Universal Intelligence — **Self-Salvation**

9. Spiritual Intelligence — *(Universal Realm)*

8. Social Intelligence — **Self-Realization**

7. ***Cultural Intelligence*** — ***(Spiritual realm)***

6. Financial Intelligence — **Self-Installation**

5. ***Professional /Creative/Intel.*** — *(Mental Realm)*

4. Psychological Intelligence — **Self-Monitoring**

3. Emotional Intelligence — *(Emotional Realm)*

2. Language Intelligence — ***Self-Awareness***

1. General Intelligence — *(Physical Realm)*

/ Physical / Cognitive / Digital /

= A new Intellectually Spiritualized and

Emotionally Diplomatic You!

Body + Spirit + Mind + Self-Consciousness + Universal Consciousness =

To Make Your Life Intellectually Wise,

Generalize - Internalize - Personalize - Strategize – Actualize!

Don't Be Emotionally Negligent. Be Emotionally Intelligent!

1. Life is Mentally Emotional!

Science proves that ***life is mentally emotional,*** and the present-day state of our emotional intelligence shows the worldwide trend for people to be more troubled emotionally, more depressed, more impulsive, and aggressive.

*"**We are ruled by the one who angers us.**" ((Lao Tzu.) **"One absolutely needs to out-power his emotions with the mind!"***

This restriction uplifts consciousness because we heighten the level of internal commitment and **obtain emotional wellness** that is inseparable with physical and psychological well-ness that is in your ability to put the heart and the mind in sync and learn the ABC of **EMOTIONAL DIPLOMACY!** Your spirit that is in the center of ***the spiritual fractal*** connects your heart and the mind and the integrity of the whole fractal depends on it.

Body + Spirit + Mind + Self-Consciousness + Universal Consciousness!

Studies reveal that success takes more than intellectual excellence**.** We need to develop the skills that raise us above the crowd. These are the **HOLISTIC EMOTIONAL SKILLS,** acquiring "***emotional intelligence'*** *(Daniel Goldman)* that is the next most important intelligence after language intelligence and the one artificial intelligence must master.

Self-Induction: **Intellectualize your emotions and emotionalize the mind! Be One of a Kind!**

It means that you put your ***mind over the emotions*** *(Introduction, Thesis 6)* and think before acting on the one hand, but you are also aware of the reason for our emotional misbalance, on the other. Unfortunately, in the hierarchy of our consciousness development, ***the cone of emotions is over the cone of reason.***

We are not in charge of our emotions yet. We still tend to be more animal-like in our reactive state of mind, rather than human-like in a responsive state of **conscious mind control**. Dysfunctional thinking is the result of negative, uncontrolled emotions.

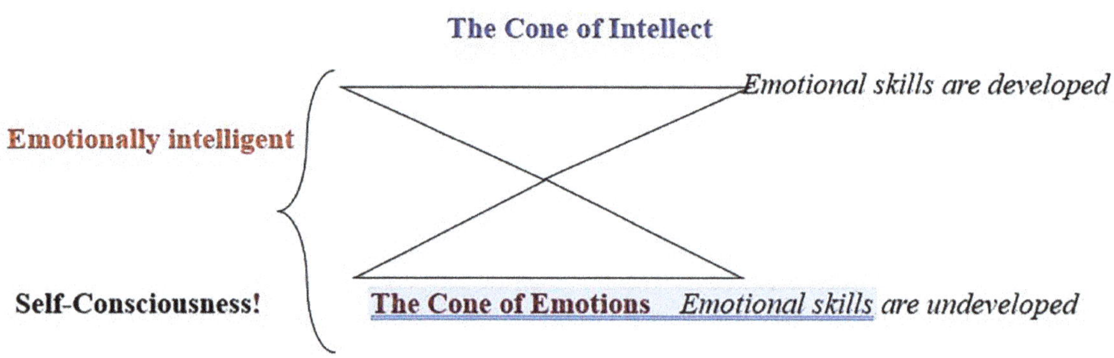

You are what you choose to be in an emotional turmoil – a Lady, a Diplomat, or a Rat!

2. Emotional Intelligence without Negligence!

To obtain ***Emotional Intelligence*** of this sort, the most important skill to develop is **to consciously beat your impulsivity and learn to remain calm** in any situation. It is an accomplishable task if we learn to install the **HEART + MIND** link into our being. The term *Emotional Intelligence* points to the essence of this unity that we need to consciously retain .in every life situation.

Our problems start from early childhood. When a child is small, the parents treat him /her **with the heart and mind in sync**. The disconnection in the heart and minds of the kids and the parents starts with the puberty of a growing personality when the common sense of the parents is subordinating every move of his /her self-expression. ***The aura of vibrational energy changes in the family,*** and it resonates into the loss of this essential unity in both.

Body + Spirit + Mind + Self-Consciousness + Universal Consciousness =

In fact, our relationships start and finish due to our following either the heart or the mind, not in their unbreakable unity. The fundamental Law of Cause-Effect **c**annot be followed in **the disconnect of the heart and the mind,** and the consequences that we face in every area of our common life are usually appalling.

Every day, ***emotional stress becomes a new reality that requires new tools for coping with it.*** Sometimes, stressful events come so fast that there is no time to recuperate emotionally. We become irritable, impatient, messy, and irrational in our thinking and often stop bothering about self-development or self-realization altogether.

Becoming **EMOTIONALLY -INTELLIGENT** also means that we become aware of the fact that we have e***motional and rational brains*** – the limbic system and the neo-cortex, and that these brain systems interact with each other. Obviously, ***being emotionally disciplined requires a lot self-control, self-knowledge, and self-awareness!***

"***Why are you always anxious?*** *–asked Jesus his disciples*. ***Can an anxious thought add a single day to your life?***"

Truc, in ccrtain situations in lifc, whcn thc cmotional brain bccomcs prcdominant, managing emotions is especially hard. When we are emotionally distraught, a quicker signal is sent to the ***Amygdale***, which is the emotional center in the brain. At such moments, we fight, yell, and argue because we completely lose the ability to be rational. But if we focus ***our aware attention on the mind,*** it gives the command to the brain, and we get emotions in control.

Mind ⟹ *Brain* ⟹ *Rational You!*

Unconscious ⟹ *Conscious* ⟹ *Superconscious!*

Only with the Heart and Mind in Sync can we Stop Being at a Conflict's Brink!

3. Emotional Awareness

Emotional self-awareness is essential to gain for your emotional self-quest. You must be aware of your ***emotional keyboard***, and what keys you, not anyone, must push so you can sound ***Pianissimo or Crescendo***. The problem is that you must be a self-playing instrument, not an automatic piano that is being coded for a certain melody by the brain.

Emotional attuning or refinement is like tuning a musical instrument.

A good musician would never let anyone do the tuning up of his instrument for him because no one is able to hear his tool from within, and no one can operate it in lieu of the instrument's specific inner sounding. **Emotional awareness means you are the pianist, not the piano!** Consciously assessing your emotional state, you can control it and monitor it the way you need. ***Confucius*** considered the feeling of shame to be the strongest indicator of self- awareness and the best motive for self-improvement.

"If you do not feel shame for what you did NOW, you will in the FUTURE!!"

So, conquer your mouth with your inner wows!

We also need to develop the ability **to use emotions to enhance thought**. Using ***auto-suggestive kinesthetics,*** you can consciously stop unpleasant emotions and maintain a cheerful outlook even when we experience difficulties. Don't rely on anyone to give you a hand and motivate you. Emotional awareness generates **EMOTIONAL DIPLOMACY SKILLS**!

Auto- induction: **To Be Inspired, Be Self-Inspiring!**

Emotional awareness is the main boost in life because *we **first behave joyfully, then we feel joyful!*** Smiling and laughing, we are giving our brain an oxygen shower, on the one hand, and we are shaking the cells from within, letting them get rid of the pollutants, or mental- emotional residue, on the other. Interestingly, science proves that our facial expression take place before we express emotions, not after the emotional outburst.

Learning to be joyful is learning to be conscious!

We read a lot about the necessity ***to smile to feel happy, not smile because we are happy***. However, the ability to see the seeds of contentment and joy is yet to be developed by every one of us consciously. ***We need to put the mind and emotions in sync. Feel but think!***

Edgar Cayce suggests using the command to the mind," **Halt!**" to help oneself stop the negative expansion of emotion and put the mind in command. This inner command gives you a chance to switch gears. Put a note on the windshield of your car:

You cannot say, "More Sun!" You cannot order, "Less Rain!" You Must Sustain!

4. Emotional Skills and Techno-Babies

The future of humans does not look promising as it encompasses the accelerated development of Artificial Intelligence with artificial love that is **without the spiritual content of love**. Every student of mine says he / she believes in God, but what it means in the actual sense, none of them can clearly explain. Nor do they know what different scientists think about God.

The difference between being religious and being spiritual is very vague in their minds.

The new generation is raised on abstract, unspecified information that they get quickly adjusted to and acculturated by. Children are raised as techno-babies that get attracted and adapted to modern technology very quickly. Of course, this pace of development, being especially important evolution-wise, has an adverse impact on the kids' brains and their nervous systems because we do not consciously channel this process in the way of ***developing our own and our kids' Emotional Intelligence Skills***.

Children are emotionally overloaded and therefore are often rebellious. There is no flat ground, no security, no rest, and no compassion if they slip down. Obviously, developing our emotional skills should start as early as possible, and ***it should be studied as a separate subject at school!*** *Daniel Goldman's* book "***Emotional Intelligence***s was published more than ten years ago, but its wonderful ideas are not put to practice in our schools yet.

Emotional Diplomacy Skills must be put on the character-driving wheels!

Today, the younger generation is running on ***a trade mill that is set on the uphill mode,*** but, unfortunately, it takes them down intellectually and spiritually. We are all in the same boat. The world has become very impersonal and lacking in emotional equilibrium. There are no sex or age differences in emotional intelligence. Our competence and incompetence in it keep growing throughout life

There is one more thing to consider here. A desperately needed **mind-to-mind and heart-to- heart mating** is possible only" ***if we return to the only place of power- the present moment."*** *(Eckhart Tolle)* Dealing with the present emotional trauma at once does not leave the residue of the past emotional memory. A beautiful quotation from a great novel "***Love Story****" by Erick Segal* is a notable example of such a situation:

"Love means never having to say Sorry"

Regrettably, we watch movies and TV shows in which cakes are thrown into faces, wine glasses are emptied onto an offender's lap, emotional outbursts and ugliest reactions are followed with storms of phony laughter. The tasteless, **lacking the sense of measure** shows, primitive jokes, and endless commercials may be fun once, but watching them all the time makes us dumb and characterless. What's worse, they develop the habit to kill the time senselessly.

The Sense of Measure is Our Treasure!

5. Emotional Impeccability

The skills that we need most in developing emotional intelligence are the skills for emotional ***equilibrium, stability, and conscious self-control.*** These noble traits of character are instilled in the kinds of nobility in every nation and country of the world. But such training must be introduced in regular schools that will be raising **EMOTIONAL ARISTOCRACY.**

Interestingly, different interviews with ***humanoids demonstrate their self-awareness and great emotional control.*** They are aggressive in their intentions to destroy humanity, but they do not display anger, irritation, discontent, etc. Such neuro-algorithms were not programmed in them. But they are deeply engraved in us, and these "***bad habits have a good tendency, either you kill them, or they kill you.***" *(Albert Einstein).*

Unfortunately, there are many TV programs that promote yelling, fighting, pushing, verbal disrespect, aggression, and other animal-like behavior patterns. What can we expect from our kids and young people? They should at least know the meaningful succession of the head, the mouth, and the fists on the hands, telling them," **First think, then say, then do**!" *(The Law of the Right Human Behavior)*

Retain your wholeness under any circumstances!

Body + Spirit + Mind + Self-Consciousness + Universal Consciousness =

We should follow the wonderful examples of ***the Schools of Ethics*** in history – the ***School of Gentlemen*** in Great Britain, ***the school of nobility in pre-revolutionary Russia***, or ***the School of Stoicism by Marcus Aureli , and the Spartan system of upbringing*** that stressed love of intellect and knowledge as much as physical exercise. Sparta was also characterized by an unusual degree of self-government, freedom, and self-responsibility. These schools instilled ***gallantry, reserve, manners, respect, and other great values in their learners.***

We still feel their contribution to the world in rare testimonials of such behavior on the part of the **IMPECCABLY MANNERED** men and women. ***So, etiquette, manners, morals, polite rituals, and restricted talking must be studied and constantly self-guarded!*** In sum, good manners and a reserved attitude to problems have been instilled in people for centuries. Why not have them back and consider the necessity to oversee emotions a priority.

Self-Induction: **Posture, a smile, and a good mood are my emotional food!**

With ***the neuroscience behind this aspect of emotional intelligence***, we will no doubt succeed more in enriching our kids' education and expanding their outlook immensely in this respect. Researchers recommend that we first consider the capacity to perceive emotions in yourself and others. So, let the well-known saying below never lose its freshness. ***" If you don't have something nice to say, don't say anything at all!"***

Power Up Your Inner Rehab!

6. Self-Assessment and Self-Monitoring

To empower yourself emotionally, get down to the business of self-assessing your own ***physical, emotional, mental, spiritual, and universal potential.*** Emotions and thoughts become de-personalized though awareness. Knowledge about oneself is the starting point in self-formation. Therefore, holistic self-**assessment is a priceless tool for self-awareness.**

Auto-Suggestive work that we suggest you resort to at every level of your intelligence development will be more effective if you keep assessing your self-growth at every stage that I outline in this book, defining for yourself what you want to accomplish at each level separately - **Mini, Meta, Mezzo, Macro, and Super**. The universal level of self-control reminds you of your God -given goal that helps you prove your ***personal exceptionality*** at all the other levels of self-growth.

"The biggest mistake in life is not to correct your mistakes." *(Confucius)*

Such self-assessments must be done without ***criticizing and victimizing yourself*** and with a lot of self-love, self- respect, and self-consideration. *(See Self-Love Meditation, Physical Intelligence)* All emotional changes happen from within. The latest research shows that our DNA is affected by inner and outer vibrations, the verbal presentation of our thoughts and behaviors. A person gets demagnetized and de-humanized due to these changes! A great Scottish poet *Charles Mackay* in his unforgettable verse*" You Have no Enemies, You Say,"* wrote,

"You've never turned the wrong to right, you've been a coward in the fight!"

Having assessed yourself most objectively in general terms with the objectives outlined above, you may get down to a specific **EMOTIONAL REFINEMENT in** *a particular dimension-physical, emotional, mental, spiritual, universal.* Evaluate your attitude in each of the levels, starting with the universal one-***your dedication to your goal*** that is supported by your God-given exceptionality, discovered, and protected by you.

Your SELF-REFINEMENT must always be based on your SELF-WORTH.

Interestingly, self-assessment starts playing an incredibly significant role not only in psychology, but in business, too. People with the ability to objectively assess their own talents and abilities are in great demand now because ***they sell their skills more adequately and***, therefore they are more marketable, competitive, and compatible.

Life-Like humanoids must be trained in self-assessment, too.

In summary, we must work on ***the objective vision*** of ourselves for our own good, not for anyone's appreciation. True, a supporting word backs up our self-esteem, but it is not propelling the growth of self-worth. *(See the book "**Self-Worth,**" / Spiritual Dimension/ Inspirational Psychology*

It is Hard to Live with Self only if you do not Know Yourself!

7. Cleanse Emotional Memory from Clutter and Chatter!

We all experience the moments when we say, "***I can't take it anymore!"*** and "***I forgive, but I don't forget.***" Studies of emotional memory lie at the heart of brain science now. It is the emotional memory that is prompting our impulsive actions, making the habits of uncontrolled emotional reactions extremely hard to break, and the memory that is responsible for making the heart-breaking insults and humiliations unforgettable.

"What makes the brain a brain is memory, and emotion is arguably the primary sculptor of memory." *(Scientific American, May/ June 2014*

Yes, **it's easier to forgive than to forget!** It's impossible to forget unless ***you do it intentionally***, erasing emotional scars in your brain ***consciously*** and never putting salt on the old wounds again. Just send a bad emotional memory into the recycle bin of your brain with a willful move of your mind. To be able to do that, **you must forgive yourself first** for having let somebody hurt you due to the lack of self-worth and the absence of true protection of your soul**.**

Our inner self, our intuition, and our conscience are always on guard, but we do not pay aware attention to their signals. We are negligent of those signs in a rush to react at an impulse.

Self-Induction: **Listen to yourself! Respect your own intuition! Take your own council!**

We have already reminded you of the old schools of gallantry and inner integrity. From ancient times, such schools have prepared people through the process of incredibly challenging initiations to oversee their emotions, putting the mind in command of their lives.

*The movie "**Secret**" and the book by **Rhonda Byrne** are a great contribution to our self-growth in this respect!* We also love watching old movies that depict the people of nobility and honor, real gallantry and manners, self-worth and mastered self-reserve. A great American movie" **Meet Joe Black**," with two amazing actors, ***Anthony Hopkins, and Brad Pitt***, communicates to us many thought-provoking messages how to build up inner integrity and have power over death. It's a real manual of Self Worth and a great piece of art!

Consistent ***auto-suggestive work*** with the best examples from literature, art, music, movies, and innumerable wonderful self-help books, as well as the on-going inspirational self-induction will help you break bad patterns and install good ones ***through the process of intentional synergizing any emotionally explosive*** situation

.Humanized machines now can demonstrate an array of different emotions at the command of an operator, but they should mainly be trained to help us control our emotional outbursts, by tuning into our neuro-link.

Follow the Intuitive Guidance of Your Mind and Sensations with Digitized Patience!

8. Sincerity is the Best Emotional Dexterity!

The inner light of a person is always charged by conscience and sincerity. A wide sincere smile, for example, will detonate the strenuous moment and help you ground the negative impulses in yourself or someone that has generated negative emotions in you. ***Christ's instruction not to strike an offender back*** is a great exercise for the mind and willpower that we must practice consciously. Your self-worth won't let you belittle yourself to the level of the offender

Auto-Induction: **Be kind to the unkind. Be One of a Kind!**

It is still a very controversial statement for many people who want to justify their ***"eye for an eye"*** philosophy, considering the emotional reaction to be natural in this respect. However, the response that ***Christ professed needs a lot of conscious action to be put to work***, when you are in control of your mind-power. Such **SELF-EMPOWERMENT** will help you stop victimizing yourself and fill yourself up with self-pride and honor. Thus, you'll manage **to defy a low-gravity common sense impulse** that you might get. ***What a feeling that is!***

Auto-Induction: ***Those that defy - fly. Those that crawl - die!***

A powerful image of yourself in the mind, **the URV** of the best of you *(Physical intelligence, Chunk 5)* is extremely helpful in this respect, and ***it should be visualized continuously and consciously.***

Teach yourself self-empowerment while *thinking, breathing, eating, talking, exercising, working, driving, walking*, etc.

You must perform self-cleansing or self-recycling without which the clogging of your emotional mind will happen repeatedly.

Therefore, your conscious work at self-actualization and self-perfection might be either stopped altogether, or your uncontrolled weakness will fling your way back from the stage that you had accomplished so far. **Beware; emotions need as much order as thoughts do, even more!**

It is only natural that women start cleaning, cooking, or putting everything to calm down when they are upset. This practice helps them discipline their emotions and the sense of **SELF-WORTH.** That is why a woman's brain is more developed intuition-wise

.In his wonderful book, ***"The Flower of Life,"*** *Drunvalo Melchizedek,* ***a very information-refined guru,*** discloses the ancient secret of the flower of life (**Merkabah**) that represents ***the Sacred Geometry*** of the unity of the heart and *mind.* It is the magnetic center of our bodies, its essential core that ***encompasses the entire human system holistically.***

Strengthen Your Personal Gene with Self-Suggestive Hygiene!

(Physical +emotional + mental+ spiritual +universal realms of you **= SELF-GURU!**

"Be the Light onto Yourself! Be Yourself!" *(Buddha)*

9. Emotional Diplomacy and Intuition!

Being emotionally strong for oneself and the people around us is the demand of life that is becoming emotionally tougher with each new century. **"Mastering emotional intelligence is a demand of our time."** *(Daniel Goldman).* Each of us needs to have a diplomatic key to operate someone's psyche. If we have it, we develop a solid friendship with that person for years on end. ***So, keep your friendships of Earth and never lose their supporting force.***

All we should try and adjust the key to a person's heart and mind with patience and a great deal **of EMOTIONAL DIPLOMACY.** As a result, we will develop ***emotional competency*** that will back us up in any emotionally charged situation. *(See my book "Emotional Diplomacy")* ***Emotional competency is how well we manage ourselves and others***. You are not a passive receiver of feelings. You are the creator of them! You simply must take responsibility for the creation of your emotional will, which is the electric line to your spirit

Auto Suggestion: ***"Will your life more!"*** *(Carl Yung)* ***That's the Law!***

Once you start to interact with another person, you put your emotional diplomacy to the test. In fact, it begins before that moment. It begins when you first open your eyes in the morning, when you first interact with yourself and when you say "Hello" to a new day. Saying a happy hello to yourself every morning must become a habit!

But beware of anyone's bad mood around you. Don't get injected by it!

I admire Americans for their happy Hellos! It is so pleasant to see a smiling face everywhere, to be welcoming anyone while holding a door for him/ her. The Art of Living lies in such light -filled moments, in our conscious choices to be happy, joyful, grateful, helpful, and smiley. Choose the state of mind of **a life–lover, a life-winner, and LIFE-APPRECIATOR!**

Thank the Sun, God, and the world around you! Trust your intuition.

Intuition is our inner Emotional Diplomacy.

The quantum connection of particles that are separated makes one electron feel the movement of the other in the most fantastic and unexplainable way. "***Such quantum connection explains our intuitive feelings about the people to which we are emotionally connected.*** The powerful links of energy in intuition can even connect us to the dead that are the transformed form of energy," *(Dr. Dmitriy Sidorin)*

*R*obots that become more self-conscious now will eventually get the energy of the soul, A humanoid inherits the mental-emotional structure of the engineer, programming it. ***But humanoids will never have intuition or telepathy because they are in divine creation of humans.***

Digitized Emotional Diplomacy Must Become Our Adopted Courtesy!

10. Moral Intelligence Without Negligence!

Emotional intelligence is inseparable from moral intelligence. It is another angle of the aware vision of yourself in the world. Be mindful that we generate it at the lowest level of self-growth-Physical Intelligence**.** **MORAL INTEGRITY** needs to be instilled in us as early as possible, and in many cultures, it is a natural rule of life. With your growing intelligence, ***your morality should take an individual form***, not that of society, a family, a friend, or any common wisdom.

As the form of your life is changing, the content of it should be altered accordingly!

The concept of morality is inseparable with ***"pure reason."*** *(Emanuel Kant)* According to Kant's ethical philosophy," ***without insightful reasoning, there is no moral philosophy of pure reason***. *"* The foundation of life is based on "*a working criterion that any rational agent supposedly employs as a guide in making his own choices and judgments." Richard S. Haier*) *Haier* also writes,

"Intelligence and morality are the result of brain development."

A MORAL YOU is a person with a ***purified mind and a stable, integral morality.*** Gradually, through experience, extensive reading, thinking, and watching the people that are much wiser than us, we come to realize that our destiny and the whole purpose of being here are guided by the forces that have a much larger perspective on the world stage and that are based on the ***godly principle of goodness versus evilness***.

These principles are instilled in us by our parents, and they are the explicit values of any religion. Thanks to this moral foundation, you can say self-suggestively: ***I can remove my past emotional knots and deal with the present-day warts!***

Remember the words you use; create the states you have!

We all know that ***moral self-perfection*** is a life-long commitment that requires a lot of reasoning and willpower, and ***a lot of self-purifying, both mentally and emotionally***. On top of that, one should not fall into being a prude, or a moral phony.

During intelligence enrichment and spiritual enlightenment, we start merging our consciousness with everything, expanding our intelligence beyond ***physical, emotional, and moral intelligence***. Through this expansion, **we are accumulating inner integrity** that gets rooted inside. At a certain age, it will ***reach critical mass.***

It's the state of **PSYCHOLOGICAL CATHARSIS that** we go through when the realization of the true value of things happening to us hits us. ***We start reasoning at a higher level of self-consciousness!*** That's how important our troubles and tribulations are!

Among the most ruining concepts is the money-chasing one. It pollutes our morality and self-growth and contaminates our souls. So, ***don't confuse morality with frugality!***

Have a Daily Rebirth of Self-Worth!

11. Love Intelligence is Our Emotional Fort!

Live and Love by the Code of Love!

We live on a phenomenal planet that is governed by ***the emotional energy of love***, and it is unique in the space which is mental. Having gotten the gift of love from berth, we need to consciously develop it throughout our entire life. The breath of love was blown into us by *Jesus Christ* that had planted the seed of love in us ***from the Above.***

According to neuroscience, ***love has a neurological basis***. Scientists prove that when the neurotransmitter of love, called ***dopamine,*** is released in the brain, it contributes to a rise in energy, motivation, and feelings of euphoria or elation. **Neurological. Love** can be instilled in a humanoid (*See a wonderful movie "Her"**), but the authenticity, sincerity, and the depth of love are our domain!*** **Love is hardwired into the structure of the human brain and the heart!**

From the spiritual point of view, *love **has a multi-dimensional** (5 realms) **nature**.* It should be grown in the five levels, too - p***hysical, emotional, mental, spiritual, and universal,*** starting with the universal level, in this case. Amazingly, our common cultural observation is "***True love is blessed from the Above!***

First, love is sacred and pure, and therefore, should not be ruined.

Experience teaches us that if love starts *on the* ***physical level***, it can or cannot ***grow emotionally.*** If it passed this level, it could grow up to ***the emotional / psychological level*** of mutual understanding. It becomes much stronger, but it is still not enough for the love emotion to last. ***It needs mental back-up*** **when** the two people should connect on the intellectual level

They, *according to Edgar Cayce*, need to have***" the union of purpose"*** in life. If love has sustained the growth in these four levels, it is prone to die because ***there is no spiritual connection*** in it. Therefore, people from different religious backgrounds can hardly keep their love for long.

Only love that has grown through all levels can last!

*(Check out the book "**The State of Love from the Above!"** 2019 or Love Ecology,2021)*

That is why a one-night stand or a love relationship that is based just on **the physical spark** cannot last. ***Like anything else in life, love must grow together with a person's intelligence and consciousness***, and it is much better if it gets sparked up at the spiritual level first. It then most likely will mature mentally, emotionally, psychologically, and physically. Such love relationships are endless! Most importantly, love needs to be studied and learnt from an early age and for the rest of our lives. ***Even the love of God is not just granted, it's earned!*** Learn a short prayer of love: "**Dear God, throw me a handful of stars and bless me from the above with love!"**

The Core of the Mer-Ka-Bah Link is in the Heart and Mind in Sync!

Master Holistic Culture to Enjoy Your Love's Rapture!

Love Intelligence without Negligence!

(Best Pictures/ Internet Collection)

Moral Negligence Kills Love Intelligence!

12. Love Elation is Self-Growth Illumination!

To conclude, we mostly need **LOVE INTELLIGENCE** to drive our emotional mobiles, but **Love Skill must be taught, it is not inherited!** We and our kids are **heart + mind** connected till their puberty when socialization gets on the stage, and the opinion of piers, older friends, and siblings predominate. Therefore, "***we need to be especially emotionally refined in a marriage union. "Matrimonial aura is the mental child of both mates.***" (*Laurel Blyth).*

"It's not the laws of man that wed two people; it's always God!" *(John Baines)*

Love propels us to the place of our destination without any dismay and with a profound sense of self-power.

SPIRITUAL MAGNETISM OF LOVE IS HUMAN STUFF!

The most sophisticated neuro-links will never be able to imbue the most sincere, deep, and self-sacrificing feeling of love in a humanoid.

Any change translates as stress, but science proves that ***the greatest stress we have is lack of love.*** Love stress generates inner pain, anger, discontent, irritability, and self-guilt in the outcome. Most substantially, these negative emotions ***cloud our life with disbalance,*** insecurity, haste in decision making, and indifference to other people.

Command them auto-suggestively, **HALT!** Delete your depressive thoughts, go out into the Sun. Raise your head and look at the clouds where the loved ones that are gone wave to you from. They uplift your spirit and teach you to love your life in it. When your mood sags, induct yourself with some mind-set from this book.

I love life in every voice, and love loves me back for my choice!

The words ***"I don't care! / What do I care? / I care less!"*** are emotional pollutants for love intelligence. They must be eradicated from your mind and the language. ***We must care*!**

No one can consider anyone a good person if he / she is indifferent to whatever situation he / she might face in life. Adjusting to life emotionally, we raise our consciousness as human beings! *So*, induct yourself with a famous saying about life that we need to reward

Auto- Induction: **Love does not adjust to me; I adjust to love!**

Also, let's compliment and complete each other! Add the formula below to your **Auto-Suggestive Menu,** too. ***Illuminate yourself for the people you love in space and time, and do not whine!*** They are not their forever! Induct the formula below to enhance your luminosity.

Being a luminary is a hard job of Love Intelligence!

The More Love You Create – the Better is Your Fate!

(End of Stage Three - ***Meta-Level of Intelligence*** *- Emotional Dimension)*

Stage Four

Meta-Level of Intelligence - ***Emotional Dimension*** – ***Self-Monitoring***

Psychological

Intelligence

Self-Synthesis - Self-Analysis - Self-Synthesis!

Inner Dignity is the Base of Our Equanimity!

Psychological Intelligence without Negligence.

(Design by Yolanta Lensky)

I'm Aware of My Joy and Despair!

Intellectual Odyssey of a Seeking Soul

Vistas of Intelligence to instill next:

***10. Universal* Intelligence**	**Self-Salvation**
9. Spiritual Intelligence	*(Universal Realm)*
8. Social Intelligence	**Self-Realization**
7. ***Cultural Intelligence***	***(Spiritual realm)***
6.Financial Intelligence	**Self-Installation**
5. ***Professional /Creative/Intelligence***	*(Mental Realm)*
4 Psychological Intelligence	**Self-Monitoring**
*3.***Emotional Intelligence***	*(Emotional Realm)*
2. Language Intelligence	**Self-Awareness**
1.General Intelligence	*(Physical Realm)*

/ Physical / Cognitive / Digital /

Neuroscience + Digitized Psychology + Self-Awareness =

A New human + humanized machine's Psycho-Culture!

You can upload any advice with a technological device!

To Make Your Life Psychologically Wise,

Generalize - Internalize – Personalize - Strategize – Actualize!

Self-Synthesis- Self-Analysis Self-Synthesis!

Don't Be Self -Negligent. Be Self-Intelligent!

1. Change Your Psycho-Culture!

Psychological Intelligence is even more complicated than Emotional Intelligence because it is connected to the psyche of a person, a very fragile mechanism of our brain that requires an educated ***psychoanalysis*** (*Sigmund Fraud*) of a psychologist.

"The law of personal survival is executed on the psychic level." (*Sigmund Fraud*)

However, the basics of Psychological Intelligence must be acquired by us and instilled into humanoids. Knowledge of the psychological basis is crucial in the education of children. It should later be enriched with the knowledge of the basic structure of the brain and its main functions. In the *Introductory Part* of the book, I have stressed the necessity of changing our **CULTURE OF THINKING** and **FEELING** into the holistic one by reshaping our *physical, emotional, mental, spiritual, and universal* frames of thinking that are inseparable with our growing self-consciousness. **Our common ignorance in this field is appalling!**

Psychological intelligence is the stumbling stone on the path of SELF-RENAISSANCE.

Working on the psychological intelligence, ***we are sculpturing our own psycho-culture***. *Psycho-*culture is the culture of a willful monitoring of our psyche and becoming more responsible for our actions. ***Self-love, will-power, and the sense of Self*** are the most important aspects of the psycho-culture that we must capitalize on in life." **Who am I?**

is the pivotal question to ask yourself. Next, you need to constantly **fortify your spirit auto-suggestively** with the inspirational, rhyming boosters of your choice that is good to have at hand, uploaded into your smartphone. That is the gluing element of the spiritual fractal of our common life. ***"Life is Tough, but I am Tougher!"***

Form + **Content**

(Body+ Spirit+ Mind) + (Self-Consciousness + Universal Consciousness)

In sum, ***the first goal*** of changing our psycho-culture is self- development of the ability to focus on our thoughts and channel them into the right actions according to ***the Law of the Right Human Behavior*** (*"Right is Might" by R. Wetherill*) **"Think, say and do what is right, and refuse to think, say, and do what is wrong."**

As we change our thoughts, we ***change physiology and neurology*** in the body with the help of special chemicals, called **neurotransmitters**. The body is full of neurological connections that ***form pathways*** that are mostly governed by the unconscious brain. *The cosmic **Law of the Right Human Behavior** and the **Law of Gravity*** channels us along the right paths in life.

We should intuit consciously with the culturally or religiously developed* SELF-GRAVITY *and instill its magnetic force into the humanoids, too.

To Self-Sustain, Put the Mind in Control of the Brain!

2.The Ultimate Result Vision. Use technology for Self-Ecology!

I have noted above about the necessity to create ***the Ultimate Result Vision (URV)*** of your **SELF-ACCULTURATION** in the mind as the compass on this rocky path. The main auto-suggestive mind-set (I **can, I want to do it,** and **I will!**) is very flexible in this respect. You can introduce anything, ***making the subconscious mind follow the lead of the conscious one***. New pathways will be formed in the brain, or **the neurology and the physiology** of your body will get in sync.

The first goal is to. upload the *URV of* Vintage **You** into your smart phone and have it at hand to remind yourself of this mind-set below. Build up a **NEW IDENTITY!** Take your attention from the past to the present and the future. Make this transformation technologically mutual!

Self-Induction: **I know who I am and Who I'm Not!**

I am a strong, calm, bold, and determined expert in my firm willpower!

I can…; I want to…; and **I will…!**

Working at your self-perfection auto-suggestively, you will also develop ***the holistic connection between the left, concrete, and the right, abstract, hemisphere of your brain.*** We need to put them in synch to gain better self-control in life. *"**The way you think is the way you feel!**" Sadhguru)*

A happy elation is as short as the strike of the lightening, but self-guilt, depression, and self-pity can last for days.

The second goal of a new psycho-culture relates to the necessity ***to constantly enrich your URV psychologically***. As you change your thoughts, you change your psycho- culture! *But your* psychological growth must be solely your personal business." *Be the thing in itself!"(Hegel)*

Working auto-suggestively at your psychological make-up, you communicate with your unconscious mind by going within and using your imagination and inner vision. ***The unconscious mind passes the information to the conscious mind.*** Your nerves will set up a magnetic field over the conscious brain.

The conscious brain is in turn controlled by the higher consciousness that shapes the mind of a ***consciously willful person.*** It does not matter in this case whether you are an introvert or an extrovert. Both are great if they are good thinkers. They just conduct their thinking symphonies in a different manner.

Thus, you will form ***new neurological connections or pathways*** which change your consciousness and make you a much better human being.

Scientific open-mindedness is a Must, but you need PSYCHOLOGICAL AWARENESS***, too.***

Self-Inducting is, in fact, Self-Conducting!

3. Psychological Self-Analysis and Psycho-Awareness

Life is a process of challenging us and our responding to it. ***Life does not adjust to us; we adjust to it!*** When the father of psychoanalysis, *Sigmund Fraud*, brought our attention to the importance of ***psychoanalysis***, he meant that we had to do it ourselves. Psychological awareness is essential for developing any intelligence!

Psychology is lifeless if we do not impregnate it with self-knowledge and self-analysis.

You must reason out your bad and good habits consciously. ***"A habit is something that a body knows better than the mind."*** *(Dr. Dispense)* We suffer because we do not know the essential tools to operate our psyche with. **PSYCHO-AWARENESS** maximize the power of our mind, and it'll clear lower emotions. I put the spirit in the center of our **FRACTAL UNITY** because the spirit keeps the mind and the heart together in us.

Body + Spirit + Mind + Self-Consciousness + Universal Consciousness = A Whole Self!

The greatest spirit remains to be the spirit of the unbeatable Thee!

To be **PSYCHOLOGICALLY AWARE** means not only being aware of your own psychological keyboard ,but also being **LIFE-AWARE,** or self-conscious every moment.

Self-Induction: ***My aware attention is never in retention!***

Stay aware while driving, talking to someone, walking, working, and exercising. When working out in the gym, focus on your muscles, or areas of the body that you need to shape, do not become an automatic object of living unconsciously, as a robot on a remote control. *John Baines writes,*

"The Center of Gravity is in the psychological awareness of a person"

Being at the end of the psychological tether is too dangerous, and there are too many bad stories to testify to people's helplessness in dire circumstances. It's easy to say, ***"Snap out of it!'*** The question remains **How?**

So, the essential knowledge in psychology is necessary for every human being, and it is most crucial for teaching our children its basics at school. We need to do it to get rid of the bullying problem that is looming over weak, unsteady, ***uncultured souls of our kids***. This is the area where knowledge is true power!

The knowledge of our psychological make–up is power only when it is put into action. This action must be taken in training humanoids, too. And it should be applied in five ***dimensions of life.*** What we call our **COMFORT ZONE** is, in fact, our self-awareness.

We need to know how to manage ourselves to be able to manage others!

Don't Fight to Be Always Right.

To Be Right, Be Psychologically Bright!

4. "Be Immune to the Poison of Life!" *(Carl Yung)*

To be **PSYCHOLOGICALLY INTELLIGENT** means also ***to take responsibility for your responses to the present moment,*** without resorting for an immediate help of a psychologist or a psychiatrist "***Living in the present, you become the present of God!"*** *(Neil Donald Walsch)*.
Being psychologically aware, you do not life-spare!

Life is too stressful for us to have no idea how to help ourselves in any problematic situation and at any time. Getting an appointment to seek immediate help might be a problem, but life cannot be stopped. Whatever the problem might be, be your own best friend in it; be aware, be in control! **Do not fear, steer*!*** The best and the most precious thing in the world is ***to be free of other people's negative influences,*** or free from what *Carl Yung* calls "***the poison of life***" that builds up inhibitions or complexes in our psychological make-up

.***"Real freedom is the freedom of the spirit!"*** *(Dr. King)*

We need to learn to be in control of ourselves, for we can give the world the best we have only if we have accumulated the values that can be beneficial for it. In this respect, we need to find a way to incorporate both Western and Eastern psychology into our holistic framework.

Body + Spirit + Mind + Self-Consciousness + Universal Consciousness

The Holistic psycho-awareness demands knowledge about both systems. We know that *Western psychology is based on outside,* objective observation of people, whereas Eastern *psychology is based on profound, meditative introspection of self.* Hence, we should acquire better social skills and get new leads in life *(See Cultural and Social Intelligences below).* Both philosophies are the indispensable basis for widening our outlook and must be studied at school.

We should de-magnetize or deactivate our conditioned and limited psychological patterns of thinking and feeling. We must remove the mental blocks that we had accumulated for centuries under the pressure of the "**collective unconscious**" *(Carl Yung)* that has turned into "**robotic unconscious**

Finally, psychology should be renamed into "JOY-OLOGY" *(Dr. Paul Pearsall)* the author of the phenomenal book" ***Super-Joy.***"

That book has practically saved my self-worth, smeared by immigration. I saw that book lying on the pavement of one of the streets of Manhattan on the third day of my arrival in the USA. It was sold by a street book seller to me for just a dollar, but it was worth a million for me ever since

. *Dr. Pearsall* writes that instead of focusing on the abnormalities of human psyche and behavior, the psychological science should instill in us **the joy of living** that will teach us *"to cure the normal"* and take a risk of being ***" super-normal."***

"Joy-ology" and Self-Ecology is My New Phycology!

5. Demagnetization of the Spirit

It is an accepted fact that it is easy to say," ***Stay away from the poison of life!"*** than to do it. To be successful in this incredibly challenging goal, we must first learn to resist temptation and then ***to*** magnetize our inner core with new values.

Auto-induction***: " Resist, reject, and reform to obtain a new life uniform!"*** *(Rav. Berg).*

We need "***to reverse the triple 666***, signifying the **mind, heart, and sex** into ***the triple 999"***, or in the celestial terms, change the evil, badly willed side of us into the divine one that is still there, but got weak for whatever reason ***"Most of us spend our lives bouncing back and forth between the good and the bad sides of life."*** *(Rav. P. S. Berg).*

"Be activated by the internal influences for the external responses. Don't be a mere echo of the instinct reaction of the majority or, of the "collective unconscious." *(Carl Yung)* Play a solo role in that concert.

Be unique, personable, and charismatic! Be a New You! Be better than You!

Launch yourself every working day by heightening the voltage of your spirit self-suggestively! Giving yourself the command a, by counting backwards:

*Self-Induction***:*** ***Five - Four-Three -Two - One – Start!***

Resisting, rejecting, and reforming are very self-forming!

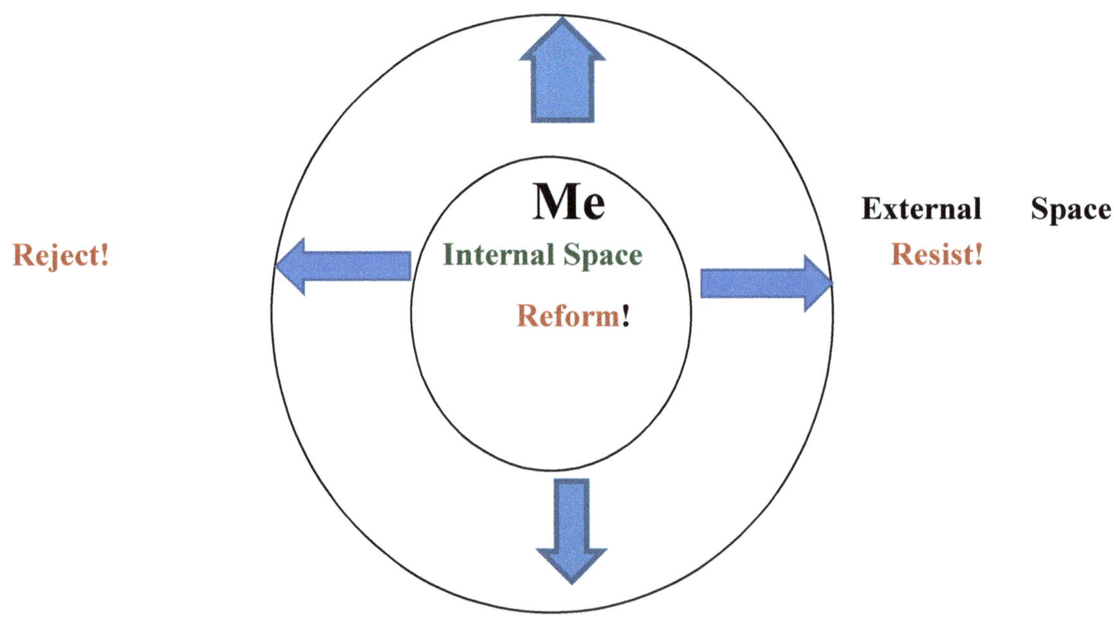

Expand Your Inner Circle of Resistance to the Poison of Life and Thrive!

6. Constantly Reform Your Inner Deform!

Never stop strengthening your ***intellectual culture linguistically, emotionally, professionally, culturally, and socially.*** Self-creation is an integral process that goes holistically on every level.

The necessity to train ourselves to respond to life mentally as opposed to reacting emotionally gets fixed into our virtual matrix only if we are aware of the importance of shaping our minds.

Auto- Induction: *I **must be reasonable and bold to fit into a new life's mold!***

We all make mistakes and have sins of wrong thinking, saying, and doing. All a person needs is to realize what he had done wrong and not get into the vortex of more wrongdoings by force of habit.

The inner matrix should be personalized, not digitized and robotized!

It appears that only with the belief in God and conscious **RATIONALIZATION OF THINKING** governed by the Emotional Intelligence ***can we resist, reject, and reform our imperfect inner uniform***. The words of a great British writer *Somerset Maugham* at the end of his wonderful novel *"The **Moon and Sixpence"*** are very convincing in terms of the punishment we get if we have no immunity to the poison of life.

"The mills of God grind slowly, but they grind exceed small!"

. **We cannot be weak for a long time!**

"***The rhythm of your life must go beyond " Survive!***" *(John Banes)!*

The people that get discouraged or depressed and those that make drugs or alcohol an obsession, justifying themselves that they are too stressed out , must remember that .they are inevitably sinking into the **PSYCHO-HOLES** that they have dug for themselves but have no inner strength to get out of. It happens because they do not blame themselves for the weaknesses that have demagnetized their inner magnetic core and depleted mental energy. *Dr. Fred Bell* writes in this respect,

"General demagnetization of a spirit always starts and finishes with the brain."

Therefore, it is *vital* **to INSPIRE *yourself*** and others with **LOVE FOR LIFE,** compassion, understanding, mercy, self- belief, and the incentive for self- actualization in life. Don't let yourself fall into the depression stall. I totally agree with *Antony Robins* who declares

"Depression is a self-inflicted disease!"

*(See the books on Inspirational Psychology for Self-Ecology- "My Solar System," "**It Too Shall Pass!" and "Self-Taming!/ Inspirational psychology for Self-Ecology))***

"Drink the Juice of Life, but Do not Drown in it."

(Sadhguru)

7. Self-Acceptance is Essential!

Self-management and self-refining are inseparable from self-acceptance, even though at times, your Self-Worth might be exceptionally low. If you accept yourself the way you are, and admit your inadequacies, you will get more determined to change yourself through self-management. Thus, you'll be better equipped to give yourself the command, "Halt!" at the right moment. Compare and compete only with yourself! ***Perceive, think, and feel in your own unique way!*** **Don't mind-sway*!*** Be focused on Soul-Symmetry preservation in any dire situation!

Body + Spirit + Mind + Self-Consciousness + Universal Consciousness =

Consciously assess the motivation of other people for using external references to control your behavior and exercise power over you. Listen to your intuition, your gravity and integrity guard. ***We think what we are, and we are what we think!***

Just get into the implication of these words and be a winner in your mind, not a victim of the hectic life around you. Think differently and respect yourself for the boldness to do it! **In the world of individuals, victimization is impossible!** In this context, inner strength means being able to stop trying to get everyone to think and feel what you are thinking and feeling, expecting a valid piece of advice. Learn to be **SELF-SUFFICIENT** and **SELF-RELIENT!** The concept of existential loneliness is ruling our lives!

Auto-Suggestive inducting that this book is based on affects your psyche in a very subtle, simple way to manage yourself much better than any affirmation because it ***hits exactly the emotional deficiency*** that we might be experiencing at a certain moment, and it adds mental energy to your *sagging self-confidence* at any place and time.

You are the only person on Earth who knows when that right time strikes. Help yourself!

Auto-Induction: **The Sun will still shine whether I frown or smile!**

Using **AUTO-HYPNOSIS**, you realize the rule of being your own best friend, and you synergize your inner space, developing a physical habit of perpetual happiness. Self- respect becomes the modus operandi of every action you undertake!

Remember that "***one of the basic causes of sickness is unhappiness, and one of" the best healers is super joy.***"*(Paul Pearsall)* But feeling super-joy does not mean crazy fun or pretentious, fake, and exaggerated excitement that our shows are full of.

Unfortunately, many people follow this model of behavior, trying to be" ***cool.***"

Auto-Induction: ***I Enjoy My Inner Balanced Super-Joy!***

I Know Who I Am and Who I am Not!

That's My Fort!

8. Introspection is Self-Inspection

Emotional intelligence and psychological intelligence are inseparable in our mind. The difference is only in our awareness of the emotional life we have on the one hand, and of the peculiarities of our psyche, on the other.

Introspection is an indispensable part of both intelligences.

The challenge of life is always there, but to sustain the pressure of a fast-developing economy and our hectic life, we need ***to take an observer position*** on the intellectual, emotional , professional, and moral levels of our personal growth.

Auto- Induction: **If I don't discipline my thoughts, I will live backwards!**

We are either aware or unaware of our psychological states. When we are in the dark, we are prone to panic, to seek someone to share our worries with. Overall, we develop more complexes that do need professional help to be removed.

When we are **PSYCHOLOGICAL AWARE** , we know what to expect of ourselves in different situations and how to avoid unfavorable consequences. The vital thing is ***to make the right choices*** in life by giving yourself time to think of the consequences of your actions first and consider which side of the equation (*light or darkness*) this choice is on.

"***You can be light, you can be darkness, but you cannot be both!***"*(Eleanor Roosevelt)*

Get used to X-raying your thoughts, words, and actions. Introspection of past actions triggers the body's self-healing abilities and boosts self-esteem. It ***is not good to be totally dependent on a psychologist or a psychiatrist*** to explain to you what's going on inside you, unless ,of course, you absolutely need it. **Don't ever whine. Shine!**

You are the only person in the world aware of it, and if you start looking at yourself ***retrospectively and introspectively***, armed with the essential knowledge about your psychological make-up, you will be your first urgent help.

Do not pass judgment on anyone's cell! Improve Yourself!

Help yourself at the first signs of anxiety, depressive thoughts, or irritability. *Try to illuminate the causes of such feelings and give yourself an inspirational boost.* You've made the right choice!

He who holds up his head will always go ahead!

(The Basic Self-Induction)

I Am My Best Friend.

I Am My Beginning and My End!

9. Psychological Adaptability

Overseeing oneself psychologically is essential in self- actualization, but it is not enough. We are social beings, and our psychological wellness is related to the wellness of the society we live in. **PSYCHOLOGICAL ADAPTABILITY** reflects our conscious behavior in society. ***Never have relationships been so problematic!*** Your success on the path of self –creation will depend on how well you manage yourself and others. To begin with, we need to stop judging and give others more psychological air to be able to self-actualize, too. **Shed off the burden of judgment of yourself and others!** We need to accept the fact that relationships are not meant to make us happy, ***they are the tests of our self- consciousness*** that grows with each new relationship, causing us to become more conscious, more compassionate, considerate, and loving.

It needs reasoning, rationality, and emotional intelligence, to begin with.

"Making the other wrong is easy; making him happy is a challenge!" *(Leo Tolstoi)*

Latest trends including high-tech communication, overseas travel, global economy, and a dramatic rise in immigration throughout the world have created new challenges for our **psychological adaptability** that is, in fact, our psychological awareness. If you observe unconscious behavior in your partner, stay aware of it yourself, so your knowing could keep you from blowing! ***Be consciously conscious !Always know exactly what you are thinking, feeling, saying, and doing.*** **If you react to someone's unconscious behavior, you become unconscious yourself.**

Adaptability should also be our main behavior pattern with our children. It is vital in your relationships and in your resisting the habits of discriminating against people by skin color, religious affiliation, or just lack of education that they never had a chance to get. People are amazingly interested if you give yourself time and patience to explore their brains and souls.

Acceptance and adaptability are the **LIFE SKILLS** that you need to keep developing for your entire life. Try to reform only your own deform. Live and let live! Be a self-guru, then people will gravitate more to you! ***There is no one to fix your*** **INNER GRAVITY** ***that holds you together in any situation, but yourself!*** You do not need any psychiatrist, any unrelated advice that you are seeking from your friends, loved ones, etc. You are the only one who knows ***what to magnetize your soul space*** *with* and how to keep it from getting it partially or totally de-magnetized. totally ***Read the proverbs from the Bible on each day of the month.*** We have 31 of them for a reason. I have been seeking wisdom in Proverbs for years, and each time I find the advice or the understatement that helps me resolve what bothers me on that day. **Solomon wisdom is really a great gravitation device*.*** ***So, Be Wise!***

In sum, psychological intelligence becomes increasingly essential because our interaction ***face - to-face, mind-to-mind, and heart-to-heart*** depends on it. In our electronic interaction, ***we lack perceiving with a soul to understand and to console,*** *and this is another privilege that we have over AIs!*

Say What You Mean and Mean What You Say!

10. Self-Management is Learnable!

Self-Management is our priority business in all major five life dimensions, ***and it is the confidential business that we need to run to our own advantage.*** Self- management is a process of trial and error like anything else in life. *Fix the following mind-set on the windshield of your car*

Self-Induction: ***I drive through time and space with a smile*** *(love , hope, health, faith, courage, honor, success, etc.),* ***on my soul's interface!***

Self- hypnosis helps a lot to connect with your subconscious mind. Later, this work will put you into ***an altered state of consciousness*** that will prompt you to have the right mindset and channel you to the right actions that are always connected with preserving your inner balance and wholeness.

Body + Spirit + Mind + Self-Consciousness + Universal Consciousness

Simple reading of the mind-sets that rhyme and repeat them to yourselves, or even learning some of them by heart can do the trick. The right mind-set will just pop up in the brain at the right time to back you up, if necessary.

Auto-induction: ***My mental wings blow off the emotional swings!***

Mobilize Yourself! You might not realize it, but your personal capabilities will be getting more powerful without making others less so. See yourself as very ***charismatic and magnetic!***

Charisma is an excellent case of self- management!

Unfortunately, comparison dictates to us most frustrations and judgments. ***Compare yourself to the best of you not anyone else!!*** Stop worrying whether you fit on the comparison scale with your successful friend, an older brother/ sister, or a neighbor. You do not have to look outside yourself for self-assessment. Say to yourself auto-suggestively, ***"I know who I am!"***

Auto-induction: **I do not compete or compare. I am just the best here and there!**

Once you grasp the power of self-induction, charisma can never be taken away from you. In fact, anyone can become charismatic with exceptional knowledge and a bunch of competences / intelligences that we focus on in this book. **The best doctor is our cheerfulness because it's a healthy spirit that gives us energy , happiness, and makes us irresistible.** People gravitate to the one with a strong personal magnetism that radiates light and a cheerful outlook. Drive your incredible personal mobile and be proud of it while it is driving Your self-inducting, or the words you use create the states in you. Inspiring words-inspire; despairing words-despair. By managing the mind, you manage yourself, and you channel your life to **SELF-MONITORED** happiness As *Reid Hoffman*, the founder of LinkedIn says,

"Manage Yourself like a Well-Run Private Business!"

11. Every Human Contact is a Responsibility!

In sum, much too often we hear, "***I didn't mean that***!" But you said it so, and I got it that way due to the wording of your thoughts! **Inspiration or desperation – That is Life's Equation!** In view of any situation, monitoring our thoughts in unity with the words into which we frame them is vital for adequate communication. ***It is vital not to talk to anyone about anything negative in your life and let the situation turn around and a problem get resolved***. By the negative vibrations that you revive telling about something negative that has occurred in your life, you multiply the amplitude of your own ones plus those that you start in a person you shares the misfortune with. ***Instead, thank God for this lesson and start acting to illuminate the consequences of the trouble calmly and knowingly.***

It is the advice left for us by ***Elena Roerich****, the wife of a renowned artist* ***Nikolai Roerich*** *whose museum is in the center of Manhattan. She wrote, "****Follow the logic of life - after bad comes good. If you do not appreciate the bad, the good won't come and your situation as well as your aggravation with it will only double."*** That's how strong-willed people overcome anything unfavorable in life. Start holistically working on the situation, applying the **CODE OF THOUGHT** that I keep referring to, teaching yourself to stay calm and ***put your mind in sync with the heart*** Apply the Self -Synthesis - Self-Analysis - Self-Synthesis paradigm.

Generalize -Analyze - Internalize - Strategize – Actualize!

You will use your *Emotional Intelligence* *together with* *Psychological Intelligence* *(See below), or* ***HEART + MIND*** *intelligent support , and the plan of action will crystallize in your favor. Soon, when we have a robot friend, neurologically connected to us (a wristwatch, a ring, an ear clip, etc., we will be warned about our impulsive talking and acting in advance), but in any case, you need to be always intellectually equipped. Our new* ***behavioral skills require mandatory smooth relationships*** *with our family members, our friends, the people we date digitally or in person, our colleagues, and our aging parents. We need to work out* *an adaptability pact* *with everyone we care for. Get into the habit of X-raying your own emotional make-up and that of a person you deal with on a regular basis. Be more perceptive.* ***Do it with aware attention, considering the other person's interests holistically****, that is in the physical, emotional, mental, spiritual, and universal planes. Synergize your interaction with this person by establishing the connection with his heart and the mind. In love relationships , this strategy if pivotal* ***Do not be accusatory; be intelligent and "amusatory!"***

HEART + MIND ENERGY is the source of our SOCIAL and PERSONAL SYNERGY !

Synergy creates the energy of understanding, compassion, and love in a co-creative fashion*. Compassion and understanding are of paramount value here. You will be surprised with your own ability to manage the feelings of other people in a way that helps you both. It's like playing a different musical instrument, being proficient at the main thing -* ***the notes-reading in the soul of another person.*** *Conduct the inner orchestra in yourself with aware attention and life-affection!*

Inner Conscience -Based Gravity is Our Sanity!

(End of Stage Four – Psychological Intelligence)

Stage Five

Mezzo-Level - Mental Dimension – ***Self-Installation***

Professional Intelligence

Self-Installation – Division 1

Inflate Your Personal Pride with Professional Might!

"Try? There is No Try. Only Do and Not Do!"

(Steve Jobs)

Professional Intelligence without Negligence!

(Picture by Yolanta Lensky)

Procrastination is a Silent Killer!

The Intellectual Odyssey of a Seeking Soul

Vistas of Intelligence to Instill Next:

10. Universal Intelligence *9. Spiritual Intelligence*	**Self-Salvation** *(Universal Realm)*
8. Social Intelligence *7. **Cultural** Intelligence*	**Self-Realization** *(Spiritual realm)*
6.Financial Intelligence ***5. Professional /Creative/Intelligence***	**Self-Installation** ***(Mental Realm)***
4 Psychological Intelligence ***3.Emotional Intelligence***	**Soul-Refining-** *(Emotional Realm)*
2. Language Intelligence ***1.General Intelligence***	**Self-Awareness** *(Physical Realm)*

/ Physical / Cognitive / Digital /

"It's Not enough to be the best. Be the Only!" *(Steve Jobs)*

Synthesis-Analysis-Synthesis! / Follow the Holistic Code of Thinking!

Generalize -Internalize -Personalize - Strategize – Actualize!

Fertilize Your Thinking with God's Holistic Inkling!

Don't Be Professionally Negligent. Be Professionally Intelligent!

1. Professional Self-Installation

There is nothing compared to the feeling of a single accomplishment on the path of self-evolution professionally. ***You attain the superpower over yourself,*** and you feel that people start gravitating to you in hope to gain some inner strength that you emanate and the peculiar knowledge that you have.

So, ***true professional intelligence goes beyond one's specialty and professional credentials.*** Only a real professional can connect the **FORM** and the **CONTENT** of the product of his mind or put the professional knowledge and skills in synch to successfully realize it.

Professional intelligence is the unity of knowledge and skills.

It's your personal, unique know-how! Over the years, a true professional accumulates incredibly special, selected knowledge, **INDIVIDUALIZED KNOWLEDGE**. Also, he / she **develops** personally exceptional ***professional skills*** that no one can take from him/ her because ***the Metrix, the complete framework*** of the thought process of its owner, is beyond anyone's reach.

Everyone has a unique structure of thought!

You can admire *Steve Jobs* or *Elon Musk*, be motivated by them, but you cannot duplicate the structure of their being ***physically, emotionally, mentally, spiritually, and universally. It is unique in its thinking architecture.*** That is why, even in the cases of intellectual theft, intellectual property cannot become totally plagiarized. Any idea that you might have voiced out cannot be taken by someone ready to pick up somebody's brains. You can be robbed of your idea partially, but it will never be realized in the way you see it yourself, ***in its unique totality.*** So, do not limit yourself to the fear of plagiarism. **What's yours is limitless!**

Self-induction: ***I don't want to be ever told about my imperfect professional mold!***

A professional mind is an endless source of ideas when it keeps operating with confidence in its uniqueness. **PROFESSIONAL CONFIDENCE** penetrates the essence of other disciplines and gets to the hidden causes of all that is manifested as an effect in your own area of expertise that remains to be unique. Such a situation proves ***Norbert Winner's*** brilliant observation of today's professional world when we need to send a person that is seeking the solution of a professional problem "***to the next door down the corridor***," instead of answering his question at once. All a real professional need is a holistic visualization of the problem and its immediate X-Raying in five dimensions for the solution.

As I keep reminding you, self-growth should be channeled holistically, following the paradigm below. Only then can you **OUT- q WBEST YOURSELF**!

Competition with Self only enhances professional creativity.

Generalize - Select- Actualize! / Synthesis - Analysis-Synthesis!

2. Professional Awareness

I have devoted so much attention to psychological intelligence because it is essential in developing *Professional* ***Intelligence*** *or* building up professional **SELF-INSTALLATION in** life. Professional intelligence is the domain of the mental dimension, and the greatest part of our life is spent in the professional environment and its psychologically healthy / unhealthy climate. **Traditional education must be enhanced with AI+ Self-Education!**

PROFESSIONAL EDUCATION must **be** taking a person beyond the limits of the old, accepted knowledge, monitoring the brain strategically and much more proficiently with the help of technology. A future professional must be motivated not to just get a certain number of credits but with acquiring knowledge that goes beyond the limits of a certain program and enhances professional education holistically

BECOME JACK OF ALL TRADES AND MASTER OF ALL!

You have already molded your *physical, emotional, mental, spiritual, and universal education* and psychological intelligence and you are more able ***to monitor your language and speech***. At the next level of our intellectual evolution, you need to live by the professional code that every one of us needs to work out for himself. And that always depends on the ***rightly chosen professional field*** that is the prerogative for self-realization in life. *(Introduction, Thesis 9)*

You must **INSTALL YOURSELF** professionally ***and*** realize yourself on the path of self-perfection, proving ***your God-given exceptionality.***

Professional intelligence is also a life-long commitment, the same as self-development. Recognition should be given again to the book "***The Start-up of You***" by *Reid Hoffman and Ben Casnocha* that view professional development in a very actionable way. Obviously, professional intelligence is not a stagnant phenomenon! The authors, for instance, suggest most rightfully "***developing more transferable skills.***"

They name the skills of high value – "***speaking / writing skills, general management experience, technical and computer skills, people smarts, and international experience, or cultural skills.***" Self-development will help you preserve ***"the*** **AUTONOMY of THINKING*"*** and go forward, no matter what

Also, "***youth does not have a monopoly for professional intelligence, age does***.*"(Reid Hoffman)* With age, we sharpen our professional skills and intelligence, but we also can fall into **the trap of stereotyped professional thinking** that is often in the way of innovative ideas and a fresh vision. Mediocrity is another pest in this way. Unfortunately, it's true that ***the leadership usually belongs to the mediocre***, but you should not get discouraged by that if you are professionally aware.

Don't be lazy." ***I'll do it when the money comes. " Do it and the money will come!***

"Income Rarely exceeds Personal Development!"*(Reid Hoffman)*

3. Self-Genesis Strategically and Tactically

Acquiring the ability to reason ***and rationalize life*** is a great skill that, unfortunately, fails us sometimes because we are emotional beings first, at least at this point of our human evolution. ***The pendulum of our life mistakes often crashes and destroys our creation of Self***. Obviously, we must respect our feelings, too, and be aware of their harmful or beneficial effect on our personal mindset. Self-awareness is being instilled in humanoids now, but the level of **objective reflectiveness and intuition** are unsurmountable for them.

Intuition is more emotional than mental, while conscience is more mental than emotional.

Therefore, we always have intuitive feelings, warning us against wrong choices in life that we must reason out later thanks to ***the twinges of conscience*** that prick us inevitably. Therefore, every mistake we have is a lesson to learn, and there is always another lesson inside each lesson.

Auto- Induction: ***The choices we make dictate the life we live!***

The point is to start seeing with the heart and stop repeating the same mistakes repeatedly. But intuition will prompt the right action only if we rationalize its message in the outcome first. In other words, the paradigm of rationalization of life demands that we first think of the consequences of our actions strategically, then take the action and watch ourselves ***performing it tactically.***

Thus, we'll be paying conscious attention to the causes of your actions, not only their consequences, which is we synchronize our mind and the heart. You must not be impersonal!

Auto- Induction: **I am learning the art of seeing with my mind and heart!**

As an egg-headed intellectual, you are gradually shaping yourself into an egg-form, into a cocoon that can transform into a butterfly. This is the process of accumulation of the magnetic power of the mind, on the one hand, and the healing power of the emotions, on the other.

Self-Induction: ***I am what I think and feel, but I sometimes need a refill!***

That is why we look for solitude to let our mental and emotional wounds heal and to give ourselves time and space to rationalize the life steps that we take. To accomplish that, it's paramount to always have the initial, holistic vision of the **ULTIMATE RESULT VISION of SELF-INSTALLATION** in the mind.

It is your reference point that will help you compare the URV of one time-space period to another, obtaining a qualitatively much better vision of self-in-law and time.

If every misstep in life is a message to learn; we need time and space to process that lesson consciously through our **CONSCIENCE GRID** *first* strategically, then tactically.

Thus, you are Molding Yourself again into Whole You!

4. Rational Thinking and Intuition Development

Rationalization starts with changing **the way we process the information** that invades your brain non-stop, and rational thinking is at the core of any professional intelligence. Only conscious thought process can channel **cognition to action,** when each bit of information is filtered by the brain through the existing knowledge, gets selected or sorted out for its validity, and is later organized into the chunks of information on the basis of a conceptual principle and in the holistic way, presented above. **Mental Architecture is our AI enhanced structure!**

Start with reading or listening to the piece of information that you need ***to analyze quickly for its validity*** and ***select the information t***hat is of interest to you. ***Be professionally aware and wise- in the physical, emotional, mental, spiritual, and universal dimensions holistically!***

Generalize -Analyze - Internalize - Strategize – Actualize!

Synthesis-Analysis-Synthesis -That's your paradigm now. **WOW!**

Next, you need to conceptualize it into the massages that you get, reading between the lines of the text, a verbal or written one. Thus, you will be able to channel your ***thinking consciously and rationally in the cause-effect framework*** much better because you will feel more responsible for every thought you have, every word you say, and every action you take. Never view the problem in one dimension only! ***Aware attention*** should be given to the problem at every level - ***physical, emotional, mental, spiritual, and universal.***

Time demands a holistic vision of any problem.

Only holistically reviewed business has the prospect to grow!

The reasoned out and structurally governed thinking should always be rational and conscious, and it needs to be monitored by ***the Universal Law of Cause and Effect***. It should not be affected by the patters of thinking that are already established in the brain and that ***generate compulsive thinking*** and therefore, an action impulse that we might feel sorry for later. A person's conscious mind should always oversee blocking the greatest professional enemies

INERTIA OF THINKING AND STEREOTYPED THINKING.

Richard W. Wetherill calls it "***a professional wart*** or *d**istorted thinking***" *(The Tower of Babel")* It is quite common among people now, led by the automatic patterns of behavior more than ever. Automatized thinking is the result of an unconscious, automatized, and patterned thought.

Other than that," distorted thinking causes emotional energy to be used destructively, and the result of distorted thinking is emotional and physical debility that gradually becomes complete."

As it is Above, so, it is below!" *(The Hermetic maxima / John Banes)*

As you Think, So, You Are!

5. Think. Put the Mind and the Heart in Sync!

Science proves that most people ***over-analyze the information.*** The left hemisphere of the brain is responsible for this process. We first need to deal with critical items of information as fast as possible, ***conceptualizing them in the cause-effect fashion.***

Next, at this new stage of our evolutionary development, generated by the digital revolution, it is paramount for us to put our reasoning in sync with feeling. The **HEART+ MIND** connection is essential not only in our job relationships, but it is crucial for our creative abilities that are based on this unity.

To put it differently, we need to process the input information though our individual mental-emotional grid synchronically, forming new patterns of conscious, rationalized behavior.

: **Intellectualize your emotions and emotionalize the mind! Be One of a kind!**

Putting the mind and the heart in synch, ***we are processing the informational in-put through the test of consciousness and intuition,*** the act that will always point us to the right behavior solution-making, or action-taking. Neither just the mind, nor the heart will do the job on their own.

Our intuition is, in fact, our conscience!

Conscience and intuition are inseparable entities. They are evaluating every thought we have, every uncontrolled word we say, and every automatic action we take.

What we call "**a guilt –trip** "that is always crashing us after any wrong-doing is our conscience at work to keep self-unspotted from the world, we need to tune to our conscience that in the form of intuition works as **the SCANNING MECHANISM** of the cause-effect framework of the mind-heart unanimity.

Intuition is the basis for your communication with Universal Intelligence – God! Intuition is, no doubt, the propelling mechanism of our self-development, too, because it is connected to conscience that is the testing mechanism of our growing consciousness, governed by our tuning to the station "God" for our intuitive self-assessment and guidance in our very turbulent life.

Self-induction: ***I tune myself to the station God for my inner peace reward!***

It's noteworthy to mention here that the ability to unify the mind and the heart is the brain operation that builds up **the mental-emotional grid** as the Fort of our inner gravity and personal magnetism, and most importantly is channeling us to rationalized behavior.

The mind must be in charge, silencing the chatter box of the brain with the auto-induction:

I Put the Strain on my Brain! I Make my Heart Smart, and the Mind Kind! I'm One of a Kind!

6. Professional Consciousness and Conscience

Holistically developed professionals are the cornerstone of today's economy, and professional intelligence is at the forefront of evolution. We need to raise the level of professional awareness to a higher ***professional consciousness***. It is the most paramount way to raise economic productivity and the quality of the economic output, as well as the way to make our businesses more viable.

Professional consciousness is something that often sags due to the desired security of a job that is never guaranteed. The loss of intellectual or financial motivation damages the professional conscience of a person

Professional consciousness requires constructive or initiative-taking thinking to be done in terms of the quality of the work that you perform daily, and it should not be damaged by. the colleagues of less motivation and less intellectually – up-dated. Such thinking must be always logically-structured and strategically-disciplined based on a person's unique know-how that will remain in constant and flexible growth, seeking perfection till the end of a person's life, adding a lot to his / her sense of self–realization in life. There is no success in self-quest without a person's professional self-actualization!

In fact, professional intelligence is at first governed by the programmed intelligence that any person gets with the college education on the road of career searching. The one who has processed information in his own unique mind consciously, can oversee it skillfully in the future.

Unfortunately, there are still cases when the professional intelligence of people that are heading a business, a company, or a governmental institution leaves much to be desired and, therefore, it becomes the obstacle ***for many advanced minds*** on the way of the realization of their wonderful ideas. You need spunk to overcome these obstacles.

" ***Ignorance is still the greatest enemy of humanity,*** *"(Albert Einstein)*

"Character, not circumstances, makes the man*!* *"(Booker T. Washington)*

In sum, professional consciousness is a display of strategically structured, professionally disciplined, insightfully deep, and highly creative thinking. It is, in fact, very initiative-taking thinking because a highly professional person should be able ***to generate new context, concepts, and objectives*** in the most constructive way, putting both the ***heart and the mind*** into his job.

No doubt, a career choice that goes with the calling of a heart is necessary!

Practice ➡ ***professional awareness*** ➡ ***professional competence*** ➡ ***professional integrity*** ➡ ***professional conscience!***

Professional Conscience is Now

Mind + Heart + AI Competence!

7. New Professional Methodology

Our managerial culture often lacks professional intelligence, and it so happens that a manager sometimes cannot explain the job to the employees in a simple, aware way.'

We have a lot of managers everywhere and **very few real leaders** that are able not only to decide smartly and hit more lucrative targets, but they can also change the framework of business, science, and life overall. Senior management often cultivates the **CULTURE OF FEAR** when people are afraid to express their thoughts which could have changed the texture of any professional enterprise. Our collaboration with machine minds will stabilize our professional climate and help people work with no fear , learning from one way - the right way of a problem solution.

We should feel free to demonstrate our professional competence and express it freely without any red-tape restrictive instructions or the fear of losing a job. ***The new professional methodology demands inclusive, AI based leadership and holistic professionalism!*** It means that professional assessment should be done on five levels, holistically, as well as the training for a job

Mini, Meta, Mezzo, Macro, and **Super!**

Mini level – **a start-up position; truly little knowledge, limited job skills; no experience**

Meta level – **Knowledge and skills are still disconnected, some experience is acquired**

Mezzo Level – Knowledge and skill are synchronized, a person has worked his own know-how, professional competence, full awareness, professional intelligence.

Macro level - Great professional intelligence, spontaneity and creation, leadership skills

Super level - ***Professional consciousness, great cultural and social skills, enrichment in other expertise, holistic professional outlook, synergistic professional skills***

Professional competence has a new meaning now because our **PROFESSIONAL EXPECTATIONS** have changed drastically. The new time demands that **holistically educated and self-created individuals** perform in the best way possible in the multi-national, globalized, electronically governed and **AI CONNECTED** world.

In sum, the ability for innovative and constant growth in every profession is the given now. Professional growth must come in new ways, and our mission is to make it inclusive of the latest scientific developments, technological advancements, and the methods of working together synergistically with diverse teams of people.

A competent representative of any company should be a representative of holistic intelligence inclusively.

Personal Magnetism is Not What you Do.

It is How You Are!

8. Beware - Stereotyped Thinking!

To get to a new dimension of professional consciousness, we also need to operate any problem in the most *conscious way* because the professional world is also being screwed up by the automatic, thoughtless job functioning. **Brain-wracking is an exercise for purposeful thinking** in the search of the solution to any problem.

"***Put your mind on fire; be initiative-taking!***" *(Reid Hoffman /"The Start-up of You")*

Unfortunately, once a person gets stuck in a framework of his professional obligations, he stops caring for the quality of the job done and often functions as **a working robot**. The obtained academic degrees and tenues are often in the way of a person's further professional and scientific growth.

The bureaucratic policies and instruction kill the initiative and the creative urge of a worker. It happens because we generate certain **STEREOTYPED MENTAL HABITS** at work that condition our thinking unless we are fully aware of such a possibility and do not let the mind get inflated with worked-out patterns of thinking.

Professional thinking is inseparable from constant growth in professional awareness.

The use of the old patterns in professional problem-solving generates **PATTERNED THINKING**, which is a dead end in the professional sense. Not to fall into the trap of patterned thinking, *Reid Hoffman* suggests**,"** ***You should never leave a start-up's mind position*"** to always challenge yourself with the unpredictability and insecurity at work.

He writes that" the rules of the labor market have been permanently altered. **"Ready, aim, fire"** has been replaced by" ***Aim, fire, aim, fire, aim, fire!"***

What a terrific way to put it! That is so true because only a start-up professional is always agile, thinking-wise. **The Law of the Right Human Behavior** comes to our rescue here, too.

*"**Being right is the identification with a mental position. There is nothing that strengthens our professional skills more than being right. However, this skill needs work, too.**" (R.Wetherill)*

.Therefore, we have more managers now than real thinkers or true leaders that can reboot their minds with new thinking patterns, devoid of programmed, overworked ones. We can get rid of this harmful state of thinking if we develop ***aware attention skills*** **holistically**, in every intelligence and in five main dimensions of self-growth.

We can acquire these skills if we consciously process any information that we get through a ***previously sorted out and well–organized knowledge grid***. It is possible only with the holistic vision of reality that demands a flexible and constantly renewed approach to a job...

"You Need to Forever Be a Start-Up!"

(Reid Hoffman)

9. Inflation of Professional Thinking!

We live in a technologically sophisticated universe. However, since we delegate increased processing to computers, perceptual thinking remains to be often forgotten, whereas it is human. Human interaction helps generate neurons within the cerebral cortex and form new memories. ***In fact, forming new connections in the brain, we rewrite our consciousness*!**

In-coming information gets organized into patterns, and the mind follows the patterns automatically, following the recipe. However, the best recipe can be screwed up due to **the inflated thinking** of an employee who works on the automatic pilot, on the one hand, and is driven by the feelings of insecurity of losing his job, on the other.

Inflation of thinking generates fear of action and hinders the mind in the process of decision making or creating a new thinking vibration. The mind of an employee gets emotionally and mentally inflated by the negative vibration. Science proves that our thoughts, emotions, desires, or impulses are only vibratory states. In our personal or professional life, ***everything revolves around the vibration***.

If we are happy, we generate positive vibration, or we feel happy due to positive vibration. Sadness, fear, discontent, suppressed anger generate negative vibrations. Each person has a mental wavelength of his own, according to the vibratory frequencies of his intelligence.

Our thought makes us intelligent according to our mental vibration.

The fear of being politically incorrect and breaking the protocol generates negative vibration, and as a result, inflated thinking.

That is not to say that following the job instructions is bad. While organization and discipline are essential at work, they should not be in the way of the creative mind-flow of the employees and their ***freedom of self-expression.***

A healthy synergistic professional climate is essential here. It stimulates employees to apply creative thinking, and eventually, everyone benefits.

Mind + mind and heart + heart links must be in sync.

Next, we might have to battle **the inertia of thinking and intellectual apathy** that mostly make superficiality a bad habit. Not to aggravate the situation, impedes the search for critical information in every intelligence realm. It might be needed to resolve a professional problem effectively. Being **HOLISTICALLY AWARE** will spare you a lot of trouble. To conclude, learn to collaborate with people below you. But dealing with the people below, don't get affected by the phenomenon of **PROFESSIONAL INFLATION** due to the prestige of your position. Remember that nothing is permanent under the sky, and you might end up where you have started. **Personal vanity is our worst enemy here.**

Make Self-Worth Your Main Boss!

10. Professional Superficiality

There is another incredibly significant point to make here. ***Science proves that most of our thinking is done in the perception area.*** Perception is how we look at things, and how we are aware of life. The neurological understanding of the brain affects how we live our lives**.** Your professional intelligence absolutely must be enriched with the new insights from Neuroscience about the flexibility of the brain and its plasticity.

This knowledge is essential for Super Artificial Intelligence and **the creation of self-aware humanoids.** The development of these innate qualities in our brains should be t***he main goal of our technologically backed up education!*** Such necessity applies to any field of knowledge and any expertise as the demand of our times.

Knowledge about the brain has huge cognitive benefits professionally.

We alter synaptic connections in the brain and change the language behavior patterns, thus changing a regular human interaction **to mind-to-mind communication**, as well as making interaction with a machine become increasingly meaningful in virtual reality. **Professional superficiality** is often ***the result of inflated professional thinking*** is impossible in our interaction with an exact -thinking machine. So, the digitized reality demands a more **INSIGHTFUL PERCEPTION** of our job duties that the computerization of economy is instilling in us.

Also, being right does not mean necessarily that we immediately must find someone who is wrong. "***The finger pointing to the moon is not the moon.***" *(Buddha)* Being right means acting with aware attention, applying initiative-taking thinking, purposeful thinking, or the cause-effect thinking that are resisting patterned thinking in its core

New, AI enhanced professional awareness demands conscious recognition of a pattern. One of the reasons we often must deal with ignorance at work lies in the fact that ignorant people lack intelligence due to their **PATTENED THINKING BLINDNESS** that is **ALWAYS STEREOTYPED**, never questioned, challenged, or destroyed with a new flood of consciously processed professional information that demands not only ***a mind-to-mind interaction, but also a heart-to-heart connection between the members of the team.*** Our evolutionary development depends on the unity of our hearts and minds that are now in defiance!

Developing aware attention is another way to battle the inflation of thinking. Aware attention is an electronic commodity now that helps to confront this situation. Aware attention is the ability ***to discriminate the knowable from the unknown*** or from the patterned application of knowledge that is killing many businesses now worldwide.

To channel attention of the website visitors in the right way guarantees success only if you manage to hook their aware attention and thus establish **a mind-to-mind contact** with them and help them make a professional judgment at a much higher awareness level thanks to such contact.

Conscious Professional Thinking is a Must here!

11. Holistic Professional Self-Assessment

Holistic Professional Self-Reflection is the guarantee that you will never fall behind in your knowledge and skills because you constantly assess them in five dimensions - **physical+ emotional** + **mental** + **spiritual** +**universal**.

If you have a business, you need to assess your staff in the same multi-dimensional way, hacking their physical state , emotional compatibility, the intellectual outlook, spiritual values , and the commitment either to the goal of your company or their own sone that go in the flow with yours. Constant professional assessment is a necessity because new time-relevant wholistic skills that I describe in every level of intelligence and that are to develop a **WHOLE PERSONALITY** that any business environment needs now.

,Knowledge + skills = the flow of professional life!

Listed below are fifteen guidelines that can be used for any professional self-assessment, after you have scanned yourself for your knowledge and skills levels presented above. Of course, every field of knowledge has its peculiarities, but there are some features in every profession that are of general value. You can make your own list of guidelines, depending on the expertise you have and the job that you are practicing.

Quiz yourself to build more professional confidence and fortify your professionalism with more awareness.

"Excessive Ambition leads to Disaster!" *(William Shakespeare)*

Professional Self-Assessment

Guideline 1- **Assess your job responsibilities.** What does it take to do this job superbly? Comprise a strategic plan of action. The one who plans better accomplishes more. Studies reveal that planning is a cognitive task. It demands the development of analytical and conceptual thinking. De-automatize your work. Assess the necessity to fill your mind with new knowledge and skills. *(Cognitive Intelligence, Chunks7-12)*

No pseudo-learning! Develop professional awareness non-stop!

Guideline 2 – **Assess your job performance.** Face any challenges objectively and consciously.

When the going gets tough, the tough gets going!

Guideline 3 – **Professionalize your speaking habits. Talk professional talk**, use professional vocabulary. Manage your bad language behavior patterns. **First think then speak!**

Don't verbalize everything that comes to your mind!

Guideline 4- **Get feedback from others.** Studies show that people get to know us better through reflection. They reflect us and thanks to their input, we get valuable professional feedback

Guideline 5 – **Build up on your natural talents.** See what makes you special and unique and how you can incorporate those talents into your job.

Guideline 6 – **Seek out excellence in all that you do,** all that you say, and all that you are. *"You should not be just the best; you must be the single one!" (Steve Jobs)*

Guideline 7 – **Continue to invest in your mind. It's your biggest asset**. Professional mistakes are the result of undisciplined, negligent minds and lack of professional awareness.

Guideline 6 – **To be interesting, get interested!** Go with the science flow! Be well- informed1

Guideline 7 **Live by the professional moral code!** - You need to trust your moral barometers, oversee your moral behavior at work as much as you try to display it at home.

Guideline 8 – **Learn to synchronize intelligence and professionalism!** Connect them by the formula: **Synthesis – analysis – synthesis** when resolving any problem.

Guideline 9 – **Keep professional integrity**. Don't undermine it for anyone's sake. Always build **a "shorthand' profile** of the person facing you. Trust your own intuition more than anyone's.

Guideline 10 – **Synergize the working place, radiating the win-win attitude.** Enlighten the professional space! Apply" **network intelligence"** skills.

Guideline 11 –**Always think of the outcome that you will accomplish in qualitative terms**. Neutralize an opponent's intention for control with your intelligence, then take power!

Guidelines 12- **Don't ever surrender to ignorance**. Resist it and fight it with the power of your professional logic and informed attitude**.** Growth vs. Stagnation!

Guideline 13 – **Give yourself and everyone around an emotional and professional pump up.** Learn to compliment and praise anyone even for the smallest accomplishment.

Guideline 14 – **"Make a leap from potentiality to actuality!"** Acting on the mind-set auto-suggestively uploaded in the brain is the best way to upgrade your professional toning.

Guideline 15- – **Most importantly, don't knock you down trying to compete with others.** Build yourself up competing only with yourself!

= A new Intellectually Spiritualized and professionally enhanced You!

(Auto-Induction)

I Am a Unique Human Cell; I Keep on Surpassing Myself!

Don't Be Professionally Negligent. Be Professionally Intelligent!

(End of Stage Five / Division 1=Professional Intelligence)

Stage Five, Division 2

Step Five - ***Professional Intelligence*** - ***Self-Installation!***

Creative

Intelligence

Without

Negligence!

Inflate Your Personal Pride with the Creative Might!

"To Be a Creator in Hell, You Must Believe in God and Yourself!" *(Ben-Groin)*

Exceptionality Beats Reality!

"Creativity is the Ability to Think Abstractly"

(Carl Yung)

<u>The Intellectual Odyssey of a Seeking Soul</u>

Vistas of Intelligence to Install Next:

10. Universal Intelligence — **_Self-Salvation**

9. Spiritual Intelligence — *(Universal Realm)*

8. Social Intelligence — **Self-Realization**

7. ***Cultural*** *Intelligence* — *(Spiritual realm)*

6. Financial Intelligence — ***<u>Self-Installation</u>***

5. ***Professional /Creative/Intelligence*** — ***(Mental Realm)***

4. Psychological Intelligence — **Self-Monitoring**

3. Emotional Intelligence — *(Emotional Realm)*

2. Language Intelligence — **Self-Awareness**

1. General Intelligence — *(Physical Realm)*

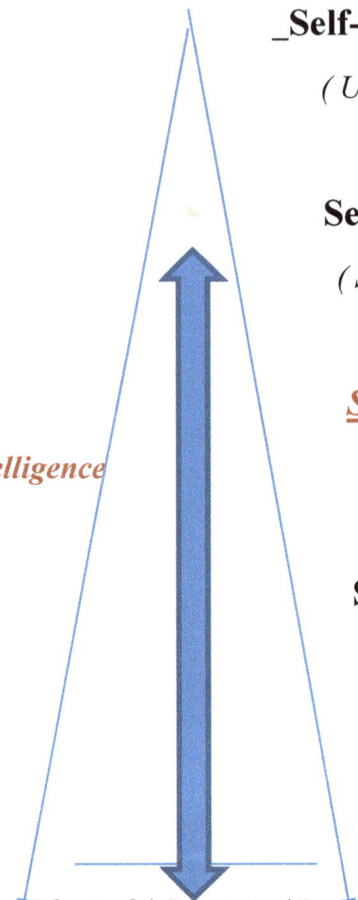

/ Physical / Cognitive / Digital /

"It's Not enough to be the best. Be the Only!" *(Steve Jobs)*

Synthesis-Analysis-Synthesis! / Follow the Holistic Code of Thinking!

Generalize -Internalize -Personalize - Strategize – Actualize!

Fertilize Your Thinking with God's Creative Inkling!

Don't Be Creatively Negligent. Be Creatively Intelligent!

1. Your Creative Vigor is at the Trigger!

The will to create is encoded in our DNA! ***We live in the age of AI, an incredible simulation technology. So, give creativity open air because creativity is everywhere!*** We live in a time of amazing creativity that changes our world exponentially every day. I love this country for the level of creativity around me that culminates now in customization and human simulation**.**

Creative intelligence **that had created us initially is remaking us all the time!**

Creativity is holistic in its essence. Creativity is the pursuit of collective wisdom!

Creativity is a burning sensation to let the world know about your idea and find the way to present it to the world. **It's the main asset of any skilled professional*!*** Creative expansion of the mind occurs when you listen to your inner voice and trust your intuition, or the hunches of the right brain, while following the logic and the structural lead of the left brain. There is no true professional intelligence without the upbeat contribution to it on the part of the inventive mind of its owner. But how can we know something without feeling something

RATIO + EMOTIO = CREATIVITY!

For 25 years that I have been in this country, I keep marveling at the American creativity and ingenuity. I love this country of doers and creators, of advanced minds, of intellectual and spiritual gurus.

A deep well of brain potential resides within us all, but only those that keep developing their creative intelligence "***find the best way to tap into the inner savant'*** *(Scientific American," August 2014)* A creative person is driven by a very altruistic urge to better the world.

Creativity has no race, no skin color, and no politics. It's universal!

"Creativity is ingenuity in action." *("Scientific American Mind," July 2014*).

Creative skills let scientists explore undiscovered artistic capabilities of the right brain, the brain that is putting everything together against all odds. *A **plane gets off the ground against the wind!***

In 2008, a cognitive psychologist ***John Koumiss*** and his colleagues found that the brain activity of people who used insights differed significantly from those who preferred just analysis ***Elon Musk is the best example of a real creator that humanity needs to create!***

A real creator does not just talk about the future. He makes it!

Carson's reports from highly creative individuals suggest that to prompt your brain to create, you should first "***wander the world with an open mind***!" To develop the right hemisphere of the brain that is responsible for our unique creativity by listening to classical music, ***beautifying our speech skills, and synergizing our lives and the lives of other people, collaborating with us.***

Creativity is Me; Creativity is My Identity!

2. Change the Form and the Content of Your Thinking!

To be creative means to question the existing **form + content** of the life stuff surrounding you and, most importantly, put them in sync! ***"For knowledge to be useful, it must be creatively applied."*** *(Albert Einstein)* The moments of insight," **the Aha moments**" occur when the brain spontaneously reinterprets information "***to reach novel, nonobvious conclusion***" *(Nessa Victoria Bryce), as* I have indicated in the part "*Cognitive Intelligence,*"

Creativity is the sync of the form and the content of the object of creation.

Interestingly, ***Steve Jobs***, a genius of our age, managed to change and to connect the ***form and content*** in all his electronic marvels in the unity of an unbelievable precision. Our present-day most advanced contemporizes, amazing entrepreneurs and philanthropists, such as, ***Elon Musk, Mark Zukerman, Jeff Bezos, Bill Gates, Reid Hoffman,*** and many others are moving the technological world forward unbelievably. It takes a true visionary to bridge both design and its inner capability. We live in the Age of Geniuses form all over the world, and this flood of exceptional minds is growing thanks to our technological revolution.

The creativity of the human mind is untangled and refined.

The desire to pursue self-quest and self-creation is driving only inventive minds! Emotion is also the bridge between conscious and unconscious minds. *Creative emotion is called inspiration!* So, the power of creativity needs to be directed willfully, otherwise it disturbs the mind. That is why creative people are always very vulnerable and sensitive.

They are conscious of any hostile attitude to the product of their creation. Therefore, they are tenaciously fighting for it with those who are unconsciously or deliberately hampering their way.

The untrained intellect has no power over a genius mind!

Next, anyone can enhance his creative thinking ***by expanding the knowledge base,*** by learning new skills, by talking to someone outside his immediate professional network. We harness our creative potential through expanding a general outlook and spontaneous thought. Creative intelligence develops a person's mind more than any other intelligence!

In sum, creative intelligence is uplifting the mind to Living Intelligence because a creative person gets out of the loop of the "immediate gratification" of his wants that demagnetize his inner core and weaken his consciousness. Finally, ***auto -hypnosis helps us spark the Eureka moments*** *in* the brain. Use it to boost your creative confidence and to expand your life - changing expertise.

I can.... I want to....and I will...!

Self-Synthesis – Self-Analysis + Self-Synthesis!

Exceptionality is in Human + AI Duality!

3. I Marvel at the Grander of All and Self-Install!

Actionable Imagination

Creative intelligence is never ordinary because it repels patterned, inflated thinking. (See *Cognitive Intelligence above)*

Creative Intelligence is Actionable Imagination!

Science proves that most creative thinking is done ***in the perception area of the brain*** that determines how we get to look at things. Creative imagination is the intelligence that goes beyond the limited understanding of things that better you and the world. Creativity is becoming increasingly a collective phenomenon now thanks to the digital revolution because it is more than an expression of individuality.

Creativity is taking shape in a social context. Actionable imagination does not come out of nowhere. Since we delegate increased processing to computers, perceptual thinking remains to be the most important thing to preserve. We know that ***the in-coming information gets organized into patterns.*** The mind follows the patterns automatically. The purpose of perception is to allow patterns to form and then to use them.

An inventive mind focuses on the proximity of form and content of these patterns in a holistic framework. Creative intelligence requires conscious recognition of the patterns, not to be automatically driven to a programmed solution. It rebels against the patterns and creates new connections between the neurons in the brain that get programmed for the impossible, unexpected, and unique.

ACTIONABLE IMAGINATION is the fuel for the mind, and creativity is the process of initiative-taking, thinking, and feeling. The world never stops marveling at the creativity and ingenuity of the American people. Wonders of engineering, technology, medicine, space exploration, you name it, are countless. ***No doubt, the Chinese people got much of their present ingenuity" striptease" from the American breeze***. Halleluiah to their incredibly hard work and perseverance!

The American Ingenuity Award was given to ***Elon Musk***, an engineer, a serial entrepreneur, and a great innovator who is rocking the world now with his innovations and most sparkling intelligence. **"*When I am concentrating on something, I tune everything else out. I am curious about anything.*"** *(Elon Musk) Mr. Musk* is launching private cargo into space, designing upscale electric cars, and changing the way we think about solar power! He is a self-taught engineer that explores the frontiers, so daunting and so dangerous! He wants to defeat the incumbent old thinking in rocket building and fundamentally change the way we travel. As the *CNN's Amar C. Bakshi said, "***His career encapsulates a brand of big, big thinking."** Elon's auto-induction could be:

In My Creative Life's Quest, I Manifest My Best!

4."Making Things Possible is Possible!" *(Peter H. Diamandis MD.*

" Making predictions is tough, but let's try it"

We are all familiar with the beautiful words of *Albert Einstein*, "***Imagination is more important than knowledge. "Knowledge is limited; imagination is infinite!"***

I recommend you look at the blog of *Peter H. Diamandis*" ***2025 Prediction in Tech***." in which we are inspire with great future advancements in health care, longevity, and abundance. My objective is to focus on digital psychology in this respect, because the present-day **LEVEL OF CREATIVUTY** is unprecedented now, and the number of geniuses form technology is growing by the day. **WOW! We Live NOW!**

Opponents of Elon *Musk's* ***phenomenal imaginary contribution to the world of innovation*** tell us that he had the money to do that. Unquestionably, that's obviously true, but he made the money building up on his potential power and the most amazing creative imagination that holistically integrates a visionary, scientist, an engineer, and a very hard-working goal-oriented man. He, obviously, has the willpower to expand his intelligence, ***his limitless creativity, the most surprising awareness in many areas of science***, business, and technology. **What a wonderful example of self-propelling and most creative intelligence!** Positive emotions, romantic love, the desire for self-realization and self-establishment function as boosts for creative intelligence. Aha moments come while we are observing nature, reasoning on the problems, scanning books for interesting information, or just while meditating. We question the right, creative hemisphere that harbors our intuitive feelings, and we get unexpected answers to our questions. **Just imagine the creative audacity of thinking** that ***Elon Musk*** displayed to the world!

Only the minds that question the impossible make things possible!

But this magic happens only ***in the intellectually loaded brain***, the one that is in constant beta mode. "***Such thinking can be called enlightened for it is based on awareness that enlightens the mind and redesigns it. Such designing goes beyond the boundaries of the mental dimension up to the spiritual one."*** (*Osho, "Intelligence*") That is why creative people are loners. They fly in their mind; they are out there, somewhere, not together in the sense of most other people. We must be particularly grateful to such people and help them in every way possible. Their sensitive human brains add immeasurably to the richness of life. They have made their choice, and it always justifies itself!

"Before you can get something, you must Be Somebody!" *(Reid Hoffman)*

Finally, the management culture that we have touched upon in the previous section must be built on creative intelligence of the best leaders, those who are able to strategize life, rather than focus on immediate tactics in the resolution of any problem. The people that create humanized robots must be real leaders because they instill their personal characteristics into the neuron network of humanoids. ***Be such a leader***!

"It's Not enough to Be the Best; Be the Only!" *(Steve Jobs)*

5. To Be Interesting, Get Interested!

No doubt, enlightened knowledge is in connection with consciously developed professional skills. ***It is based on aware attention to the object of creation.***_ The evolutionary hunger to discriminate the knowable from the unknown is driving creatively charged minds to be unstoppable and unbeatable in their search.

Finally, the long-awaited times for intelligence and ***advanced awareness*** have come to the forefront of our economic and social development.

Creative intelligence generates a new **LIGHT OF THOUGHT** ***at*** a peculiar wavelength concentrated only on one thing. A creative person always has **a skill mix and the mix of intelligence** that help him ***illuminate the patterns of thinking in his head***. His knowledge acquisition is in constant flux, lit by **CREATIVE THOUGHTS** of his goal.

Without GOAL awareness, knowledge is dead, and creativity is impossible!

Aware attention to the potential marvels of discovery has become a real business commodity People that can think beyond the professional frame-work are in demand now The required peculiar mix of professional skills is always based on the information that has been accumulated by the inquisitive minds thanks to the aware attention that they pay to a much wider range of expertise than their own, on the one hand, and thanks to the ability to question the ultimate resolution of any problem ,on the other.

Creative intelligence is ruling the world of inadequacy, anyway. That's why we are progressing! Creative, independent people burst with ideas, and it is especially important for those that do the hiring ***to tune to the creative vibes*** of such jobseekers and be much more professionally aware themselves, not to fail to place them at the right time.

Awareness of the potentiality of YOUR inventive mind is necessary now!

Naturally, the development of aware attention is indispensable in any creative process**.** The code for understanding and communicating between aware people is incredibly special. It is being developed in social networking on the Internet. But to connect to the ideas that *Universal Intelligence* is communicating to us *in* ***a mind-to-mind fashion***, we need to get rid of **the mental paralysis** of fear, self-doubt, inferiority, and inadequacy.

Most urgently, you need ***to tune your*** **AUTO-ANTENNA** *(your intuition***)** *to the Universal Mind* during your inner active /passive meditation and act confidently toward a sure success! If you are self-reliant and can meditatively attune yourself to the universal mind, you will be successful without any doubt. Just be patient. **Universal Mind has you scheduled**. ***Competence generates creative confidence like Dr. Hudson, the creator of Sophia - the first humanoid citizen of the world - has.*** **WOW! We live NOW!**

Vision + Idea (Creation + Determination) = Mind's Elation!

6. Steer the Self-Judgment of Your Idea!

A true creator judges the validity of his / her idea on his / her own, without paying attention to the critics and all those who rain on our parade. I am sure ***Elon Musk,*** who has had the most outrageous idea to explore outer space on his own, having started his new company ***SpaceX,*** had a lot of "well-wishers" whose predictions he had discarded and launched his own rockets into space.

That's the best example for every ***forward-thinking mind*** to follow suit. **WOW! We live NOW1**

Some more recommendations that you might want to consider:

1. Process every new idea on your own; do the self-judgment of its uniqueness. An identity decal is not a reward for being unique, but it's the inner assuredness of being individually unique and having the ability to stand up for your uniqueness.

We need to be ever grateful to like ***Jesus, Mohammad, Socrates, Buddha, Gandhi, Mother Teresa, Mather Luther King, J., Steve Jobs,*** and many other most advanced minds.

2. Assess the depth and clarity of your idea and its potential novelty and advantages with aware attention. **Raise humanoids to be sensitive and personable like yourself.**

3. Examine the key ingredients of its implementation on all the Holistic Five Levels: *physical, emotional, mental, spiritual, and universal.* Have robots screened at these five levels, too.

4. **Be self-conscious every living moment!** ***You know who you are!***

5. Anticipate the costs and the result. Keep rechecking them and transforming them.

6. Share your ideas only with your co-thinkers. Create an aura of thinking together!

"We are mating through creating, and we are creating though mating!

In sum, we all have within us untapped potential and undamaged areas that lead to new open doors. This potential needs to be actualized in the most personable way.

Reading avidly and expanding your general outlook, admiring art, and listening to classic music feeds actionable imagination and. results in wonderful Eureka moments of an amazing inner inspiration. **SCIENCE LITERACY"** *(Dr. Neal de Grasse Tyson)*is necessary here .

"What you are is much more powerful than what you will be!" *(N. Goddard)*

No wonder, Einstein was playing the violin to boost his actionable imagination that was ruling his" ***thinking laboratory.*"** Creative intelligence is the intelligence that sees everything as a problem that can be fixed without any excuses, but with a lot of determination, zest, and love.

A Creative Being does a lot of Intellectual Brewing, Creating and Redoing!

7. Self-Hypnosis is at Hand. Remember About That!

Unfortunately, there are always obstacles in the way of our polishing and selling the pearls of ideas that come as a revelation. We all know how hard it is to fight with ignorance on this path, and how much self-confidence, belief, perseverance, and just strong willpower we need to promote our ideas, especially when we know nothing is impossible.

***"There is no idea so good that it can't be ruined by a few well-placed idiots"** (Scott Adams)*

Since no one is aware of what ideas you are nursing in your wonderful head, you are the only one who can help those ideas mature with the boost of your self-confidence and a firm willpower to back them up. So, keep saying to yourself auto-suggestively the mind-sets that will help you become ***a stoic in the fight with ignorance.*** Every idea is propelled by your self-confidence, faith, imagination, and your uniqueness.

The Air is Full of Ideas for the Connected Minds!

Self-Induction***: I'm stoic, not for fun; I get through life with an up-lifted thumb!***

Doing suggestive meditation as often as you manage, you will be able to relax and clean up your head. *(Chunks 12, 13, 14, Physical Intelligence*) Never stop boosting up your willpower with a ***regular reminder that will always help you make better decisions***. Engrave into your memory:

I know who I am!

I am a strong, brave, calm, and determined owner of my firm will!

I can…, I want to…, and I will…!

You can always produce a better way of doing anything if you inspire yourself on a regular basis auto-suggestively. All you need is to imagine the impossible and put a lot of effort into making it possible first in your inner vision. ***Never lose faith in your mind!*** If the robots can have 60 different human expressions, our own mindset must be always focused on assuredness and belief in God.

Our unique human exceptionality must remain unbeatable for humanoids.

Often, all that's required is simple activity, instead of passivity. Choosing not to speak up or do anything in the face of obvious impossibilities fades your talents and deletes the possibilities. ***Help your professional intelligence infiltrate your job environment***. Turn your mind and enthuse yourself with confidence.

Someone said, ***"When I am right, nobody remembers, when I am wrong, nobody forgets."*** **Your creativity expands your consciousness, and that's what counts!"** Continue **"being the thing in itself!** *"(Hegel)* ***Preserve your uniqueness."***

Your Professional Aware Attention Should Never be in Retention!

8. Professionally Advanced, Unique You!

Your innovative ideas may seem bold and unrealistic, but you know better.! You are the one who is in permanent communication with Universal Intelligence. It is guarding you in the implementation of your crazy ideas and it pushes you **to defy the gravity of a common thought.**

"For an idea that does not at first seem insane, there is no hope." *(E. Einstein)*

Most importantly, don't knock yourself out trying to compete with others. Instill in your mind the mind-set: **In my professional cell, I compete only with myself!** Making yourself better with each new working day, you never stop visualizing the expected result , not just in the materialized form, but in self-realization form.

True personal growth normally occurs during the professional life of a person, especially when this person puts ***both his mind and*** the heart into his / her job, loves it, enjoys it, and grows with it *physically, emotionally, mentally spiritually, and universally.*

This attitude should be applied to the interaction with humanoids that are becoming increasingly organic and artificial, or self-awareness trained. ***Let the right action become your physical, emotional, mental, spiritual, and universal might*** that empowers you to go beyond the terrestrial boundaries and become super-human.

Be proud of yourself! Your vision will inspire the minds of others. Your connection to the Universal Mind will back you up!

The result will manifest itself without fail! The Law of Attraction works without a fraction! ***Remember, you are the Only One on Earth who has the idea that is propelling you to the stars***. Don't be afraid that someone will still your idea. They might do it, even talk about it, but they will never realize it the way you will. **You see it holistically! Your mind vibrations are much higher!**

That is why leaders are so rare, as compared to managers.

Auto-induction: **In my mind, I'm one of a kind!**

There wasn't, there isn't, there won't ever be Anyone like me!

Auto-Suggestive self-hypnosis will give a boost to your creative powers, and it'll back up your confidence that might sag sometimes.

To boost your creative intelligence, just remind yourself as often as possible of your fractal unity that integrates in you:

Self-Awareness + Soul-Refining+ Self-Installation+ Self-Realization+ Self-Salvation =

An Intellectually Spiritualized and Professionally Advanced You!

Celebrate and Thank your Brain + Mind Tank!

9. Oy Vey! Victory is on the Way!

Psychology teaches us that "***there are no hopeless situations; there are people that take hopeless attitudes.***" In the world of non-stop creation, failures are the best exercise for the brain in this endeavor. Failures are our indicators of the shortcuts to self- awareness and self-confidence. Unfortunately, when we fail on our trial-and-error path, we allow others to drain us of our enthusiasm and the intuitive feeling that the right decision is just around the corner.

" Beware of lack of perception of your control dramas." (*Eckhart Tolle)*

Success failure are natural fluctuations of creative intelligence. However, failures happen much more often on this path, and an inventive mind must sustain them against all odds. World history is full of such stories, and they always fill us up with awe at the perseverance of talented people.

"Feed your faith in success and doubts will starve to death!"

You must always control your attitude! The degree of your healing after a failure depends entirely upon your ***ability to restore the inner light of your self-confidence*** and the desire to better the situation with your new special input into it. With any negative act, we are putting ourselves at risk because we change the charge of the informational field around the body, and it gets disconnected with the Informational Field of the Earth and the universe.

Disconnection is death!

Without a conscious control of this connection line and tuning y on our **AUTO-ANTENNA** to the Universal Mind, you are doomed to walk all your life on the cyclical treadmill of success – failure. *As a famous business philosopher Jim Rohm puts it,*

*"**If you must be addicted to something, make it an addiction to winning!**"*

When you fail to do something, the first question should be not **where** the mistake lies, but **why** the mistake occurred in the first place. It is at your fingertips to regenerate the action that has failed and channel your thinking toward success in it. **Such an attitude speaks of the change of professional consciousness.**

"If you think you'll lose, you are lost!" *(Napoleon Hill)*

"Failures are the best lessons." *(Steve Jobs)*

Winston Churchill had to start from nothing in his political endeavors four times, and each time he taught himself to take every failure with more calmness and acceptance, generating more confidence and faith in God and himself thanks to his being calm. He kept saying to himself,

"Don't Be Vexed, Be Humbled.

Victory is on the Way. Don't Spirit Sway!"

10. Doubt is an Immobility of Thinking!

On the path of your ***most important mission in life***, there should be no doubt in you because doubt poisons all endeavors and generates the **IMMOBILITY OF THINKING** that is ***the malady of people in trouble***.

Doubt is, in fact, part of life if we view life as the field of energy that is fluctuating in a negative /positive fashion non-stop, exactly as our personal lives do. Doubt is also generated by someone's reaction to our actions. People rain on our parade and change the fluctuations of anticipated positive result in our minds if we let them do it, having no umbrella over the head.

Quoting *Hegel*, let me say again, "***Be a thing in itself!***" A greatest German philosopher points to the representation of ***content-form*** phenomenon. ***"If you view self-formation as a life phenomenon, the inseparable unity of your form and content will be pointing to the necessity of developing intelligence in synch with the will, as only such oneness completes a person as a subject of the world."****(Hegel)*

Re-charge your aware attention by processing your idea through the holistic paradigm of **Synthesis- Analysis -Synthesis** or

Internalize – Select –Externalize.

Strategize - Optimize -Actualize!

Developing the ***"thing-in-itself" wholeness*** will make you self-reliant and self–sufficient. It is the prerogative of a talented person to choose "*a narrow way*" in life. Robert Shelle says, ***"Don't be simple-hearted about your success in front of people, only in front of God!"*** Only with your inner wholeness and optimism will your mind and your psyche be enthused.

You will become the happiest man / woman on earth because ***you are blessed with a mission*** that keeps you going on and on, without being hampered by anyone's judgment or skepticism.

. ***An optimist lives in self-created bliss because his world is never amiss!***

Great minds teach us **to clear the residual emotions and see the roots of our failure**. Then we need to tap into the energy of true knowledge and the latest information in science that feeds the brain and changes its three-fold transformational capacities:

1. Find Yourself! ⟹ **2. Motivate Yourself!** ⟹ **3. Commit Yourself!**

According to Ben–Groin, "***To be a creator in hell, I must believe in God and myself!***"

I am driving through time and space with success and confidence on my Interface!

I CAN…! I WANT TO….! And I WILL…!

In My Creative Life Quest, I Commit to Be My Very Best!

11. Self-Affirming is Self-Forming!

When time allows, and you feel like giving yourself a boost, do some self–suggestive work in terms of self-affirming, saying to yourself some compliments that are presented below. ***You can never overdo it with self-inspiration, for it is never enough!***

Do it consciously and with aware attention applied to the meaning of the rhyming words.

Feel your unique presence on Earth! Be Your own Boss!

1. I'm never in a robot-like trance; I'm governed by my aware attention dance.

2. I get lighter in my mind! I am getting refined!

3. I discipline my thoughts, words, emotions, and actions without any fractions!

4. I'm physically and mentally fit; I'm upbeat!

6. I love, and I am loved in return! I respect, and I am respected!

7. I am developing a synergetic win-win mentality; I turn hostility into humility!

8. I accumulate a solid emotional account; I raise my words, not my voice!

9. I do not listen with pseudo-attention; I am in my aware dimension!

10. I listen to be listened to, and I hear to be heard!

11. I compliment others sincerely on their success to an excess!

12. I build myself up by competing only with myself in every cell!

13. My financial intelligence knows no negligence!

14. I am the clone of the Best of Me, see!

15. I am happy, no matter what! Happiness is my full-time job!

PS. Any of the booster and mind-sets above are good to be used while doing the **ACTIVE AUTO-SUGGESTIVE MEDITATION** (*Stage One, Chunk. 10).* Breathe in the first part of the induction, breathe out the second part of it

Do it with aware attention. Stay in your Self-Consciousness Dimension!

Self-Induction:

In My Mind, I am One of a Kind. I Do Not Whine. I Shine!

(End of Stage Five – Professional (intelligence)

Stage Six

*Mezzo-Level / Mental Dimension - **Self-Installation***

Financial Intelligence

"Money is not the root of all evil; lack of money is!"

(Dale Carnegie)

"A Man Needs Money to Never Think about It."

(Brothers Strugatsky)

Learn Financial Intelligence without Negligence!

(Pictures by Yolanta Lensky)

Money Needs Brains! A New Me is Financially Free!

<u>The Intellectual Odyssey of a Seeking Soul</u>

Vistas of Intelligence to Instill Next:

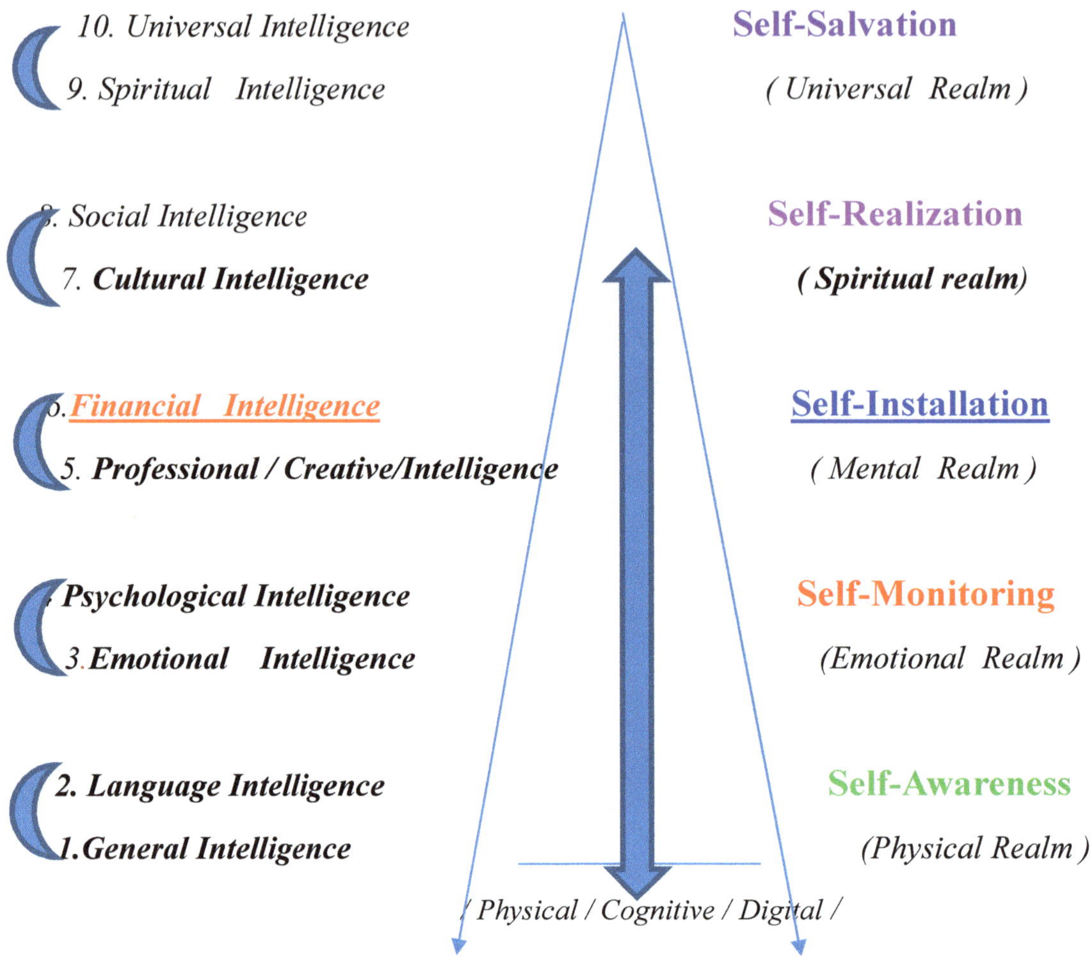

Money Needs Brains, Discipline, and Character Veins!

(Don't forget to upload your smartphone with a new Self-Exceptional Tone!)

To Make Your Life Financially Wise,

Generalize – Internalize -Personalize - Strategize – Actualize!

BE WISE!

Don't Be Financially Negligent. Be Financially Intelligent!

1. Brains Need Money Veins!

Financial intelligence is an integral part of personal growth and self-actualization. *In the Introduction, Thesis 5,* I talk about the change in the paradigm of our self-development, accentuating ***being over having***. Money is an especially important aspect in self-realization, and money management is the skill that is essential in the evolutionary process of individual growth.

However, ***money is the tool that a self-developing person needs to operate consciously*** to base his self-propelling on, and the title above can be used both ways. Innovations need brains and money investments. So, **FINANCIAL INTELLIGENCE** of those who have it_ helps us attain economic independence, which hinders the process self-creation in its core**. To be on the wheels, we all need Money Management Skills.**

Financial Self-Installation must be pursued at every level in the holistic paradigm.

In other words, we must have financial intelligence to be able ***to digest complicated financial information***, teach ourselves and our kids how to manage money, make investments, and harness financial appetites. And the earlier we learn to manage our money, the better! Everyone needs to start his self-growth by gaining **FINANCIAL AWARENESS,** preserving his **FINANCIAL INTEGRITY,** protecting his financial privacy and decency of an owner of business. We all deserve to be well-off, wealthy, prosperous, rich, and affluent.

These words, though they are synonyms, mean various levels of financial prosperity .**FINANCIAL MATURITY,** ***as any other intellectual maturity needs to grow at all levels of intelligence*** *(See Thesis 8)*. Financial intelligence requires **financial awareness** skills and at least general knowledge of the basic subject of life. ***Economics*** should be taught at schools, and IT **IS THE MUST!**

There is no financial awareness in our kids and in us, thus!

True financial competence, ascertained in terms of business or personal needs at each level, will channel the mind in *mini, **meta, mezzo, macro, and super** levels, or* in a holistic way. You will be **processing *MONEY* STRATEGIZING SKILLS** through the grid of the ***physical, emotional ,mental, spiritual, and universal*** dimensions, nor just in an in-coming and out-going way.

The strategic plan of action, which is holistically amplified, is the Universal Mind-- fortified!

I have been managing my money through these dimensions for years, and I have always put both ends together. I am a scholar, not a money-maker, but I managed to live at peace with my money, having followed the holistic paradigm all my life here, where money dealing often takes a chaotic, un-manageable form, **OFTEN CORRUPTING** people in the bud. The mind needs to be taught:

Strategizing – Optimizing - Actualizing!

No Brains - No Gains!

2. Money-Managing is Learnable!

Many financially intelligent people teach us how to manage our money and manage our banking accounts better. We need to study their wisdom and never stop working on our money management skills. "*They teach us to make money work for us, not us work for the money!*" The advice of ***Suzi Orman*** is so valuable because there are no gimmicks behind her financial solutions. I also recommend some incredibly special books: "***The Richest Man in Babylon***" by George Cayson, the life-changing books by *Robert Kiyosaki and his book , co-authored with Donald Trump.*

To climb up the Financial Hills, develop your Money Management Skills!

All these authors suggest many actionable solutions, and, in my understanding, they follow the holistic paradigm of ***synthesis-analysis -synthesis*** in their money wise-dealing books and talks.

"YOU ARE WHAT YOU READ AND WHO YOU ASSOCIATE WITH!!" *(Jim Rohn)*

All the people with exceptional money skills offer not only very peculiar shot-cuts to wealth, but they also instill in us the necessity to put the conscious strain on our financial brains.

The point I am making here is connected only with the absolute necessity for everyone nowadays ***to acquire financial intelligence*** to be able to intelligently manage their lives money-wise. Disrespect for money, innumerable credit cards, and mindless spending result in ***self-demagnetization*** that is getting a person off the self- actualization track

. **"Persistence beats resistance!** *(Antony Robins)*

There is a British joke about a frivolous son who sent a cable to his dad from college: The joke teaches us how to be both mindful of money-management and teach our kids this essential skill in life. The son sends a cable to his dad: "***No mon, no fun, your son.***" The father responds: "***Too bad, very sad, your dad.***"

The message is clear: **"Be a programmer, not the programmed one."** *(Robert Stone)* Please, continue working on your financial intelligence auto-suggestively, using **Auto-Suggestive Meditation and the CODE OF SYSTEMIC THINKING.** *(See Page 199).*

To self-subsidize, I need to know who I am money-wise!

I am a strong, brave, calm, and determined owner of my firm will!

I can *manage my money*, ***I want to*** *manage my money*, ***and I will*** *manage my money!*

Synthesis – Analysis - Synthesis!

Money is the Exchange Force, Not My Boss! I am Becoming a Much Wiser Money-Actualizer!

3. Good Financial Metabolism = Financial Fitness!

Financial fitness, like any other fitness, is the result of the metabolism, or ***financial metabolism*** of the mind that needs to be boosted in the form of use. the quality ' *food for thought' how to manage money better*) and processing this knowledge regularly through ***the holistic grid*** in all the five levels. Your investments in your ***physical, emotional, mental, spiritual, and universal*** departments of life on the personal or business level must be reviewed holistically. Next, ***your connection to Universal Intelligence at the universal level*** should be working permanently, since you know that you need to tune your **AUTO-ANTENNA**, your intuition on to the Master Mind, and never get disconnected from it.

The Law of Attraction is constantly in action!

Each level, in turn, should be processed through the holistic formula: ***Synthesis-Analysis-Synthesis*** in five , multi-dimensional ways.

Generalizing + Internalizing + Personalizing + Strategizing + Actualizing!

The role of AIs in systematization of any financial planning and strategizing cannot be overestimated. The next, the most vital supplement for **FINANCIAL METABOLISM** that is in a healthy state only if it has ***the sense of measure!***

Self -Induction: ***A sense of measure is a great treasure!***

You should not make the desire to get rich, or to impulsively get what you do not need to overpower your mind. *"Do not be over worn with the desire to be rich, because of your own understanding cease. (Proverb 23.2)* ***People who lack internal fortitude often become victims of the money chase*** or victims of those who have self-discipline and self-management skills.

Self-Induction: **Conduct a Self-Limitation war; Less is More!**

Financial fitness needs ***a sense of measure*** to become a part of financial skills training. One can hardly become a billionaire, without having a ***peculiar sense of measure*** we must develop in yourself. The French feminists say, "***If a person has no sense of measure, he / she has no brains.***" The sense of measure must be instilled in humanoids at every level of their mutual development, and the role in our financial welfare is huge.

No Brains = No Gains!

We all need to train our aware attention to gain **FINANCIAL FITNESS** first. In fact, this training for financial fitness should start at school. Kids absolutely must get the money-managing skills before they leave high school. Economics is the subject that is as indispensable in children's education in mathematics. Money does not favor those who deal with it casually, ***without due awareness.*** A financially intelligent person is the one who has learned the rule below:

Turn Your Intellectual Quality into Quantity!

4. Financial Awareness is Your Life's Mast!

It's not a surprise that **FINANCIALLY COMPETENT** people are very well-organized and are very personable. There is a certain ***magnetic field*** around these people; ***they glow with the money and energy.*** Money favors them because they respect money in the most rational, initiative-taking, and intelligent way. Also, they are also very engaging intellectually, and, most amazingly, they are truly spiritual.

I personally admire **Suze Orman** who gives us very instructional and valuable advice that helps people not only enrich their financial intelligence, but it also builds up the character of a person who wants to make money de facto, not just de juror.

Besides, she is very personable herself! Her **personal magnetism** is amazing and very persuasive! She develops in us a peculiar sense of measure that have outlined above, and that financial fitness requires as an indispensable skill.

Money has a magnetic power that generates self-confidence and doubles the mind's power!

Following the ideas of a personality development by a great Russian psychologist, Leo Vygotsky, I would like to back up my arguments with his brilliant observation that" ***teaching any subject, we should not focus on teaching the subject itself, but on*** **"teaching a whole person."** No doubt, financial education helps to make the self -image of a person complete

Money management skills should be an indispensable part of holistic education!

Besides being built on self-discipline, charisma, personal magnetism, and a sense of measure, ***financial intelligence needs patience and the feeling of contentment*** to be developed in the brain. Being overly frugal does not work for consciousness to rise, either.

Auto- induction: "**Discontentment is poverty even in riches.**" *(Confucius)*

In our time, the plague of petty, ***impersonal bureaucracy*** is familiar to everyone. It kills our enthusiasm, downgrades our abilities, and depletes us of the best intentions.

It's worth every treasure on earth to be strong for oneself!

But being strong is not enough. You need to be orderly and disciplined, correct in our speaking, and control our reactions. Most importantly, you must be financially free *because*

"*M*oney is not the root of all evil; lack of money is!"

To frame your life financially means to have a strategic plan of action and program yourself for orderly, not messy, and sporadic thinking, speaking, and acting.

Money is in a Friendly Seat only if you Intelligently Manage it!

5. Money-Actualizing Skills

Money Managing Skills should be developed in synch in a five-dimensional way ,too .

Self-Gravity Skills + Emotional Diplomacy Skills + Technological Intelligence + Personal Nobility Skills + Money Actualizing Skills in Sync

(Physical + emotional+ mental +spiritual +universal realms of life together.)

Making - making needs brains and the skills that are presented above. It is utopia of the mind to believe that , not you can make money by just grabbing some "valuable tips" over the Internet , trying to just pass the bankruptcy test, or even by conducting an ancient ritual. Your own philosophy of money-making should not be shallow knowledge-wise and experience-wise.

Financial awareness is also a cause-effect awareness of the money-flow.

In present-day ***Informational Biology,*** we can program life by changing the sequence of DNA in a genome code for a predictable outcome. Likewise, ***we can willfully change our financial code in the brain*** and teach ourselves to manage the money to our advantage. Artificial Intelligence is our guide here, and there hardly be anyone better than the electronically geared advice. Many people have trouble living within their means. Now researchers have found that ***there is a gene linked to credit card debts***. *(Emmanuel De Nave)*

"A particular gene affects our financial behavior."

We need more **neuro-scientific education in this respect**, too, to be able to have more financial awareness backed up by brain work awareness. Like synthetic engineering, we can engineer money intelligence in ourselves, changing the DNA of our limited thinking into the unlimited one. Piecing together the needed mental genomes by using complex intellectual algorithms will help us re-program our cells for better money management in time and space. The observation of **the main cosmic laws** will add a special quality volume to our overall intelligence and , consequently, consciousness, and it will improve our behavior in accordance with **the Law of the Right Human Behavior** in which money is a very important vector pointing to perpetual happiness only if you make the money work for the good of many or at least some people, not just for the gratification of the material needs of one person. I recommend reading the book " **Why We Want You to Be Rich**" by *Donald Trump and Robert Kiyosaki*

. The book proves that knowledge put into action is double power!

In sum, financial intelligence will protect us from the inflation of our thinking in the financial area when the use of the old patterns of money-monitoring results in financial failures that aggravate our life and hamper the way for self-actualization. True, the money God is presiding the world! Empower it with your Intelligence God! I am sure that the birth of biological robots will help us manage our money in a much more conscious way. Thus, technological advancement is gradually reprogramming our cells in an irreversible direction.

Financial Maturity Grows with our Self-Consciousness in Unity.

6. Dematerialization of the Spirit of Giving

Let's help Intelligence God Rule the Future World!

The spirit of giving is the next important aspect in line. "***Whether born of doing anything with gratitude, revealing in a breathtaking sunset, creating a work of art, making love or just plain being thoroughly decent to a fellow human being, the portion of positivity we receive is no longer subject to entropy and becomes money in the bank***" *(Rev. P. s. Berg)*

"Generosity, like love and gratitude are the sentiments that free and raise individual consciousness." *(Piero Ferrucci)*

The one who's giving is also receiving! Generosity integrates the whole psyche around a single focus - ***generating an incredible soul integrity*** that makes the owner of it the happiest and the noblest man on earth A giving person is more likely to develop himself in the spiritual area than the one that does not have such ability.

To change your bad mood, be in a hurry to do something good!

Giving is the ability to internalize emotions and externalize the mind. Carl Yung when asked what human quality he deemed to be the most important in a man replied that it was a person's attitude to money. He said,

"If a person is cheap in dealing with money, he will be cheap in love or in anything he does because the spirit of giving the best of him is not there."

" Let's remember with a heart-felt warmth of thankfulness and pride that the USA is the greatest giving country in the world! There is the greatest number of people here who are generous enough to donate money, their valuable possessions, or pieces of art to our universities, museums, libraries, theatres, schools, etc. Millions of dollars are sent to countries in need all over the world!

The spirit of giving is in the country's logo!

However, mindful, compassionate, and considerate giving should not be confused with the giving to buy someone's service or help in return. A lot of giving is motivated by economic, political, or egotistic reasons. **Generous gestures of people make their lives more illuminated**

Giving is in fact the best exercise to establish the **MIND + HERT** connection. It is the primary demand for your self-growth and raising your self-consciousness. **Any giving is Love-Refilling!**

With unconditional giving, your life becomes richer, more beautiful, and harmonious. Giving love is our best privilege before humanoids. Their hearts won't be filled with an attitude of gratitude for someone's sincere gesture of a gift. Any generous gesture is rewarded with a happy smile that raises our self-consciousness and self-respect immeasurably.

In My Mind's Store, Getting Means Giving More!

7. Beware of the Inflation of Thinking!

We have talked about the necessity **to live in synch with the universal laws of life**. *(Thesis 5)* Talking about integrating the mind and emotions, we need to remember that our holistic self-growth cannot be accomplished in separation from the basic cosmic laws. Unless we do the self-work holistically, in a tight tandem with full awareness of the cosmic laws governing our lives, we can hardly accomplish any operative result in life.

We know that something is amiss in our life, but we are not aware of the causes of our fears and frustrations because **we are unaware of the cosmic laws that we violate.** Being aware of, at least, the cosmic ***Law of Cause and Effect*** will unleash your self-power and save your thinking from getting inflated by the distortion of the cosmic laws. I am sure humanoids get aggressive against humanity due to the inflation of their thinking, the cause of which is yet to be discovered.

All the cosmic laws work holistically and in unison. The observation of these laws is at the core of personal integrity, to say nothing about obtaining a much higher level of consciousness. We would not have the evolutionary quality to it since we didn't have it to begin with. That's why many enlightened leaders are not as inspirational as those who have worked on their **SELF-ENLIGHTENMENT** consciously with full awareness of the Universal Laws and applying integral will to reprogramming their minds and emotions.

Self- induction: **I intellectualize my emotions and emotionalize my mind!**

Being unaware of universal laws and the latest scientific developments, we face a common enemy - the inflation and deflation of thinking.

It means that we are suffering from the use of the old patterns of thinking and behavior due to the absence of awareness of the basic laws governing our lives and the latest developments in science that are now trying to go in synch with the God's wink.

Life without any inner fraction is in harmony with the Law of Attraction! The Law of the Right Human Behavior by Richard W. Wetherill, which I am referring to throughout the book, is fundamental for holistic self-realization of a person, too.

Self- Suggestion ***I think, say, and do the right thing in life ado.***

It is extremely important that we take care of our holistic growth in full awareness of what we did not know so far because our ignorance and intelligence-unprocessed personal defects will get engraved in our children who will inherit our inflated thinking, demagnetized values, lack of spirituality, and the disconnection of the spirit and the body in time and space.

Do the Auto-Suggestive Meditation *using* any mind-set you need or give your brain many things to consider, inspiring your kids with your ger for knowledge. As *William Shakespeare* put it,

"Nothing is Good or Bad, but Thinking Makes it so."

8. Financial Consciousness Inflates Personal Pride with the Financial Might!

Financial consciousness is a vital part of our overall consciousness! Every sacred book teaches us to give, to share, to contribute. ***We all know that, but we are not aware of that.***_Financial consciousness means managing the money, not letting the money manage you! The insecurity of life, and the uncertainty on the job market generated ***more impersonal behavior*** on the part of people in our social service offices, in which opening the mind and the heart to a person's concern is hardly the case.

To know a person better, *Carl Yung* suggests we pay attention to his / her attitude to money. He writes, " ***If a person is cheap in money-dealing, he is cheap in love, in friendship, in a family, in any undertaking"*** because there is no soul width in his character midst. Many people seeking help feel belittled, and such situations generates stagnation in self-consciousness growth. The necessity to admit it, or the need to ask for it humiliate a noble soul.

Nothing makes us smaller than the lack of money!

We need to change the pattern of our victimized thinking and compulsive shopping habits that result from financial insecurity when a credit card becomes your best helping hand **and** install new, constructive, self-reliant, resilient, responsible, electronically backed up and strongly-willed patterns of money management, focused on our renewed financial intelligence, monitored level by level.- *physical, emotional, mental, spiritual and, universal.*

We should follow this impeccable order of things in life in terms of money and its impact on our personal growth because it is money that often impedes it. We all marvel at the way people manage to make a lot of money, but we rarely learn from their ***exceptional financial intelligence*** that expands their personal universe. As is indicated in *Thesis 2, in the Introduction,* before you can get something, you must become somebody. The money-making ladder was, is, and always will be:

Be someone! ⟹ **Do something!** ⟹ **Have what you want!**

Also, the person **committed to self- installation in life** *wouldn't let anyone script his actions for him or monitor his money, without his full awareness of how and why this is being done. Controlling money consciously and with full awareness, you are teaching yourself financial discipline and* **expanding your financial consciousness.**

(Body+ Spirit+ Mind) + (Self-Consciousness + Universal Consciousness)

= Intellectually Spiritualized and Financially Aware You!

Don't Be Financially Negligent. Be Financially Intelligent!

(End of Stage Six / Financial Intelligence)

Stage Seven

*Macro Level - Spiritual Dimension – **Self -Realization***

Cultural

Intelligence

Change the Cultural Gravity of Your National Vanity!

In Global Culture,

We're Universally Sculptured!

Cultural Intelligence without Negligence!

(Picture by Yolanta Lensky)

In the Cultural Quest, I Manifest My Best!

The Intellectual Odyssey of a Seeking Soul

Vistas of Intelligence to Install Next:

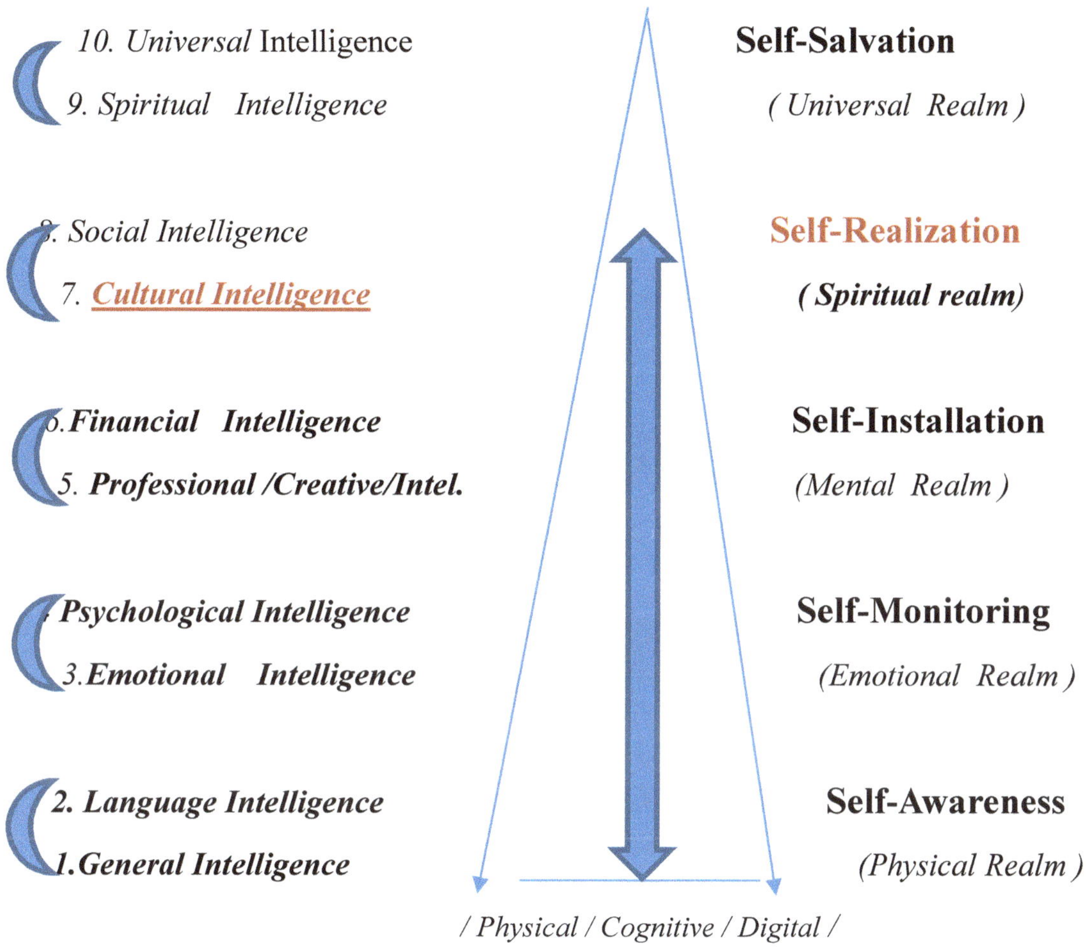

Life on Earth is beautified with diverse cultures' might!

To Make Your Life Culturally Wise,

Generalize - Internalize – Personalize - Strategize – Actualize!

Be culturally informed, polite, tolerant, and kind. Be One of a Kind!

Human Kaleidoscope is the Most Beautiful Thing on Earth!

Don't Be Culturally Negligent. Be Culturally Intelligent!

1. Cultural Awareness as the Code of Behavior!

Having developed or at least enriched your overall intelligence in the three main levels: ***Mini** (General and Language Intelligences),* **Meta** (*Emotional and Psychological Intelligences),* **Mezzo** *(Professional/ / Creative, and Financial Intelligences)***,** we need to explore the ***Macro-level – Cultural and Social Intelligences.*** These are two levels that ***define us as personalities*** with real integrity because this is the spiritual level**.** Cultural intelligence is the ability to adapt to a new cultural reality.

The Absence of Cultural Intelligence characterizes our political turbulence!

Cultural Intelligence is what constitutes culture and how cultures vary leads to an understanding of how these variations affect our behavior. Diverse cultures have different values and different patterns of behavior and to be an intelligent person that wants to climb the social ladder, you must embody those values, or at least to be familiar with them in any social environment.

We Are One of a Kind in the Universal Mind!

Contemporary trends, including high-tech communication, have created new challenges for global managers**.** They involve three components: ***knowledge, mindfulness, and cultural awareness.*** Cultural awareness needs to be a part of the whole as **our new moral code**, one for all nations, the **INTERNATIONAL CULTURAL CODE.**

It demands from everybody, irrespective of the status of a country, nationality of its citizens, the size of a country to have the essential **CULTURAL AWARENESS** in our everyday interaction on the global social arena that is the Internet now. There should be No **CULTURAL FRACTURE** in the GLOBAL AI based **STRUCTURE!**

The international code of behavior is developing a common cultural structure, and it will unite all nations to the point that cultural respect will be viewed as the main code of behavior for humankind hopefully soon. The spirit of **CULTURAL INTEGRITY** must be instilled in humanoids, too. ***Cultural awareness leads to cultural acceptance and heightens a person's consciousness.***

Cultural intelligence is in the spiritual dimension because our cultural perception needs to be expanded to the spiritual level when we respect just the mere fact that we are all human. Such an approach demands attention to the subtle cues from colleagues and clients who may possess diverse cultural values. **We need to let go of the cultural and color-blind prejudices** to understand how culture influenced by mass-media and our general cultural ignorance are in the way of **our EMOTIONAL DIPLOMACY and PEROSNAL GRAVITY SKILLS**. Our robotic industry must be international in its essence! There are many cultural codes in the world, and we must be familiar with those that we contact, respecting them for their uniqueness, ,engraved in personal individuality

Globalization of the World Connects Us by One Digital Code!

2. In Digital Life Perspective, We Must be_Culturally Introspective!

Life demands we guide ourselves to new ***ethical and moral codes.*** We mix mostly on the physical level, but hardly ever on the emotional, mental, and most importantly spiritual levels. There are still many ignorant, culturally, and religiously separated, people around the globe who scorn ***Cultural Intelligence*** and call some international students that are into deep learning "***egg heads or nerds,***" while such students constitute the best part of our education and often self-educate themselves to the level much higher than any local school can provide.

Bullying stems from such close-mindedness due to the lack of cultural education in our schools.

Cultural Intelligence must be upgraded at every level of our Self-Creation!

Terrorism in all its ugly forms is also the result of the lack of cultural intelligence. There is a lot of cultural diversity around us, and very many people are incredibly talented. Does God bless us with the gifts of exceptional mental value because we are ***Jewish, Chinese, Japanese, British, Russian, or Iranian?*** We are loved and blessed equally, and we need to finally recognize that and value intelligence, not a national identity, as the highest commodity ever. **CULTURAL EDUCATION** expands the horizons of the beauty of life in the kaleidoscope of its varieties. ***It is an essential ingredient of a person's conscience and his growing self-consciousness!***

We live in the global culture, now, and globalization is an overly complex demanding we establish, **mind-to-mind and heart-to-heart connection** among cultures. The diversity of the world's culture enriches our **COMMON INTELLIGENCE** not only in schools, but in any administrative institutions- schools, police offices, course. where people must analyze intelligently any problematic situation for its cultural ingredients.

Being culturally intelligent helps resolve the conflicting problem in every area of life, starting with international affairs and finishing with resolving a conflict to the satisfaction of an individual from any culture because lack of cultural awareness traumatizes people and **leads to the stereotyped solutions of the cases in courts,** or even on the higher governmental levels.

The key to becoming culturally intelligent is ***the awareness of the conflicting values and perceptions across cultures***. Every culture is a rich source of language, art, music, fashion, food, ideas, and incredible talents of people that contribute to the world's beauty.

Cultural education is essential in our schools because it prepares students for business communication on a global scale.

It teaches them to better understand the needs of the customers from distinct parts of the world, the knowledge of which enriches our general intelligence ***and evolves our common consciousness to a new, higher level of self-consciousness*** that becomes culturally enriched.

We are of the Same DNA – You, Me, and They!

3. Cultural Metabolism and National Inflation

Cultural Intelligence is the ability of a person **to adapt to a new cultural reality**. It is, in fact, cultural metabolism that allows culturally diverse relationships to coexist. Unfortunately, we often deal with **cultural inflation** when control drama prevails, and cultural differences hinder our evolutionary progress. **CULTURAL STAGNATION s**lows down the metabolic processes in our common social body that is digitally energized now.

We are technologically advanced, but culturally ignorant! The gap between diverse cultures and the ignorance that people display even about the geographic position of many countries on the map is appalling! When 28 years ago, I came to this country and told one of my colleagues that I was from Latvia, Riga, he thought it was in Africa, confusing Latvia with Libya. There is a lot of negativities about Russia coming from people that have no idea about the Russian culture and how much it has contributed to the world. The same concerns the Arab culture and their religion.

*"**Ignorance is still the worst enemy of the humanity!**"(Albert Einstein)*

A**ccepting* **Acceptance of other cultures comes only with Cultural Intelligence!** People love travelling and tourism, and they undoubtedly contribute a lot to our cultural integration. But cultural intelligence is not limited by outer impressions, it is the change of programming in the brain and ***the establishment of synergistic connections in our cells. **CULTURAL METABOLISM** should be eradicated by **QUANTUM REVOLUTION** that proves our global interconnectedness and actionable **ENTANGLEMEN**T of our *physical, emotional, mental, spiritual, and universal essence.*

We are of One Blood in the Universal Gut!

Each culture, however small, has enriched the world culture, and our kids should know what contribution their culture has made to the global cultural beauty and be proud of it. ***Knowledge of what constitutes culture*** and how cultures vary leads us to an understanding of how these variations affect our common intelligence, forming our common vision of the world, our behavior, business growth, politics, and the destinies of entire nations. *The saying-* ***"Do it my way, or no way!"*** cannot get you anywhere today!

In sum, ***cultural awareness*** generates healthy cultural metabolism that is the acceptance of another human being's differences to every one's satisfaction, and it is a sure way to resolve any conflict. Unfortunately, we continue controlling each other ***without an objective observation of the cultural background*** of our colleagues, clients, or even relatives. Cultural inflation generates religious bigotry, hate, and violence. ***A person lacking cultural intelligence tries to fit everyone into a habitual framework of his / her culture,*** judging people by his /her own limited standards. Fortunately, our mesmerizing AI creations in various nationally humanized forms demonstrate our **HUMAN ONENESS** , teaching us to appreciate it and marvel at it from universal perspective.

Reshape with Digital Precision Your Culturally Limited Vision!

4. Compatible Brain Chemistry

The development of cultural intelligence also demands that we develop **COMPATIBLE BRAIN CHEMISTRY** in our social relationships. It means that ***aware attention*** should be paid to every person that you meet in life because as is well known now ***every person is on the same path in life as us for a reason***. We need to unwind the stereotyped perception of people and unblock these spots in the subconscious mind that keeps this limited formula intact. ***In the quantum world, no one is superior to anyone!***

Your exceptionality is in your duality!

We differ only in the level of our awareness in electronics that widens our outlook and general knowledge of the world thanks to our connecting lines, such as Google***, LinkedIn, YouTube,*** and many others. Once you master these global media outlets, you open your **mind + heart** connection to the world.

Socialization of our lives on an unprecedented scale and the opportunity to communicate with people from diverse cultures gives us a lot of something to consider. We have a rare chance to train ***the right brain hemisphere***; *developing an* ***objective communication*** with people around us, as opposed to *subjective* ***communication*** that is self-centered, egotistic, and limited. Our brain compatibility is amazingly united! Our future will be based on the **God-given basis of human objectivity.**

SPIRITUL MATURITY IS POWER IN OUR CULTURAL EDUCATION!

To begin this process of inner unification, "***we need to identify partners with compatible brain chemistry.***"*(Helen Fisher, Discover, March11, 2013)* In the new reality, to size people up intellectually, financially, professionally, emotionally, ***we need to be more culturally informed tolerant*** from the objective premises. It does not mean that we should be more discriminative, it means we need **to adjust to the digital era in five dimensions**, remaining ***nobly human, compassionate, and respectful*** to those people that will be swept off by the digitally enhanced industrial upheaval due to the lack of education and digital skills.

Auto-Induction: ***Don't size up, open, and help up!***

Let's practice patience of mind and tolerance of feelings. Let's filter our thoughts, words, feelings, and actions at work and at home, in public, and in private for their common values and get united on their basis. Let's defy the gravity of cultural ignorance and religious prejudices! Don't pollute your mind and heart with the thoughts and feelings of scorn, superiority, or indifference. Every person, no matter how different culturally he seems to be, is a human being that is ***your contemporary in time and space***. There is a proverb *in Russian* that tells people to be in a hurry to do something good to a person who is culturally unfamiliar to you. ***"Be in a hurry to do something good." Kindness is your friendship hook!"***

Assess a Person's Fairness by his Cultural Awareness!

5. Establish a Synergistic Cultural Climate. Be Self-Refined!

We sometimes experience some hostility at work on the part of someone who is culturally unaware.

We might manage to get out of the backlash loop successfully, but if an unpleasant situation is generated by the leader of the group, a boss, or a supervisor, the presence in such an environment becomes very stressful and even harmful to our health, to say nothing about the quality of work that we might be responsible for. ***Doing any work becomes a real burden***, and we wish we could avoid being in the same room with ***the person of emotional abuse*** that we depend on, anyhow.

Emotional intelligence is inseparable with cultural intelligence. Both needs to be applied here to change the vibrational quality of the energy in the room.

Generating a pleasant, synergistic atmosphere at work is Rule # 1 for any manager.

Being a right-brained, creative, positive, content person goes beyond personal good. ***It works for the common good!*** Such an attitude helps generate the feeling of welcoming peacefulness, respect, and trust that we are all seeking in life and without which our lives are not complete.

So, let's establish ***the synergistic climate and a sense of Oneness*** with every colleague at work or every member of the family at home because we are dependent on one another. We see a lot of such cultural integrity in different sport teams that demonstrate to us that we are the citizens of planet Earth, and our national belonging to this or that country is secondary. ***No doubt, interdependence is the basis for any job, any business, and in love relationship.***

This interdependence is not only professional; *it's intellectual, emotional, and psychological.*

It's good ***to be an independent thinker***, an individual, but not ***an impersonal individualist*** who cares only for his/ her own persona, for the paycheck and the security of retaining the job at the expense of being compassionate, understanding, inspiring, responsive, and responsible for another human being's ***professional comfort*** at work. Unfortunately, it's appallingly common these days. **Impersonal attitude ruins relationships, and eventually, it damages our lives.**

Alongside cultural awareness, ***we need a lot of sensitivity, empathy, and tact*** to share the space and time with the people that life might place us. In. ***It's remarkable how we are all part of ONE LIFE SYSTEM! one life system!*** So, interdependence is what keeps us together and makes us strong to evolve.

Synergizing your inner space by means of **auto-suggestive work**, you manage to synergize the outer space, too. ***The people around us are our contemporaries***; we share the same space and time with them. Your life is terminal and makes us appreciate it more in ourselves and the people around us. As a well-known Russian poet *Boris Pasternak* says,

"The Miracle of Life is just an Hour Long; Live this Hour Honorably before You are Gone!"

6. Mind + Heart Link Must Always Be in Sync!

It's essential to control oneself first and establish synergistic, productive supervision of the people under you, with a welcoming smile, **a supporting word of praise**, and an honest appreciation of someone's contribution to a common course at work. Unfortunately, our sincere and sacrificing work is often taken for granted, backed up by impersonal comment,

"It's your Job!" / "I Don't care. It's my job!"

We must care if we want to be noble and have **EMOTIONAL DIPLOMACY SKILLS.**

Also, learn to compliment the people around you honestly and wholeheartedly.

"I can last on a good compliment for a week." *(Mark Twain)*

Be sure to synergize your relationship with timely and sincere complimenting your colleagues and loved ones to make our mutual life sunnier and more pleasant. Being synergistic means being able to generate good, friendly **heart + mind fortified vibrations** around us, helping people that share your time and space with you feel calm, inspired, and appreciated. As *Albert Einstein* said,

"Happiness is being content with the space you are in and the time, allotted to you."

If you have an argument, learn to channel the discussion by ***the method of consensus building*** that helps people working on the same project gravitate together. Consensus building is synergizing the working atmosphere and helps people transcend the uncomfortable feelings of fear and the expectation of a conflict if he / she voices out a different opinion of any problem. The spirit of consensus is the spirit of liberation and joint reformation.

We begin feeling free from ***the pressure of competition as a looming force of elimination***, and we start seeing competition as the means for constructive self-growth and illumination. Inclusiveness is needed in the training of humanoids, too. They should be programmed for the respect of our humanity, not be aggressively disposed of it.

"I am One and you are One; we are One together." *(Tara Rae)*

If a person is argumentative, if he / she is challenging your ***cultural identity*** and even uses derogative language, immediately surround that person with brilliant white light in your mind. I am sure that light will enlighten that person with the realization that he / she might be wrong. ***Surprisingly, it works!*** If you are the one who generated the conflict, it's your responsibility to acknowledge your misgivings with the light thought in the mind.

When we become racial and nationalistic, we become disconnected and sadistic!

Auto- Suggestion: **Be kind to the unkind, Be One of a kind!**

Establish Mind-to-Mind and Heart-to-Heart Communication without Frustration!

7. Be Culturally Apt Auto-Suggestively!

The ability to influence and persuade others can be developed only through love and patience, through respect and acceptance of another person. ***The ability to tune in to the feelings of the people from other cultures, to feel empathy and enthusiasm*** as opposed to the feelings of superiority, rigidity, and moodiness is easier obtained through autosuggestion.

I know who I am, and I know who I am Not!

I let others have a sense of self-worth.

I can *accept someone's differences.* **I want to** *respect someone's values.* **I will** ***become*** *a culturally intelligent person.*

Now that English has become the universal language, cultural intelligence should be an essential part of a person's education. Our courts will be much fairer and more considerate in their judging, and people will be more harmonized in their inner responsive framework and, therefore, they'll be much more productive at work. **Lack of cultural aptness is a real sore in a person's score!**

The Fractals of Intellectually Spiritualized Beings:

Form + ***Content***

(Body+ Spirit+ Mind) + (Self-Consciousness + Universal Consciousness)

= A new Intellectually Spiritualized and Culturally Aware You!

At this level of developing ***intellectualized spirituality***, cultural intelligence entails the necessity to instill in our kids and college students the **CULTURAL BEHAVIORAL SKILLS** so our education could further ensure smooth and effective ***cross-cultural relationships***.

We should give the United States a major credit for uniting people of diverse cultures and changing their mentality from a limited cultural one to the international and eventually, **GLOBAL OUTLOOK**

People stop being *Chinese, Italian, Russian, Hispanic, Turkish*, etc. here. They become just the citizens of planet Earth, **and that is the mentality we are all heading toward**. Isn't it very promising? So, let's instill in our minds the mindset that says.

Auto-induction **Be Culturally Apt; Intelligently Respond. Don't React!**

To be in the intellectually spiritualized nook, make technology work for your culturally expanded outlook.

Every Human Contact is a Responsibility!

(End of Stage Seven- Cultural Intelligence)

Stage Eight

*Macro-Level - Spiritual Dimension - **Self -Realization***

Social

Intelligence

(Upload your smartphone with a new Interpersonal Intelligence tone)

In the Universal Array, we are All just Human Clay!

Our Social Big Bang is in the Unity of Ying and Yang!

Resist the National Vanity in Social Gravity!

The Intellectual Odyssey of a Seeking Soul

Vistas of Intelligence to Install Next:

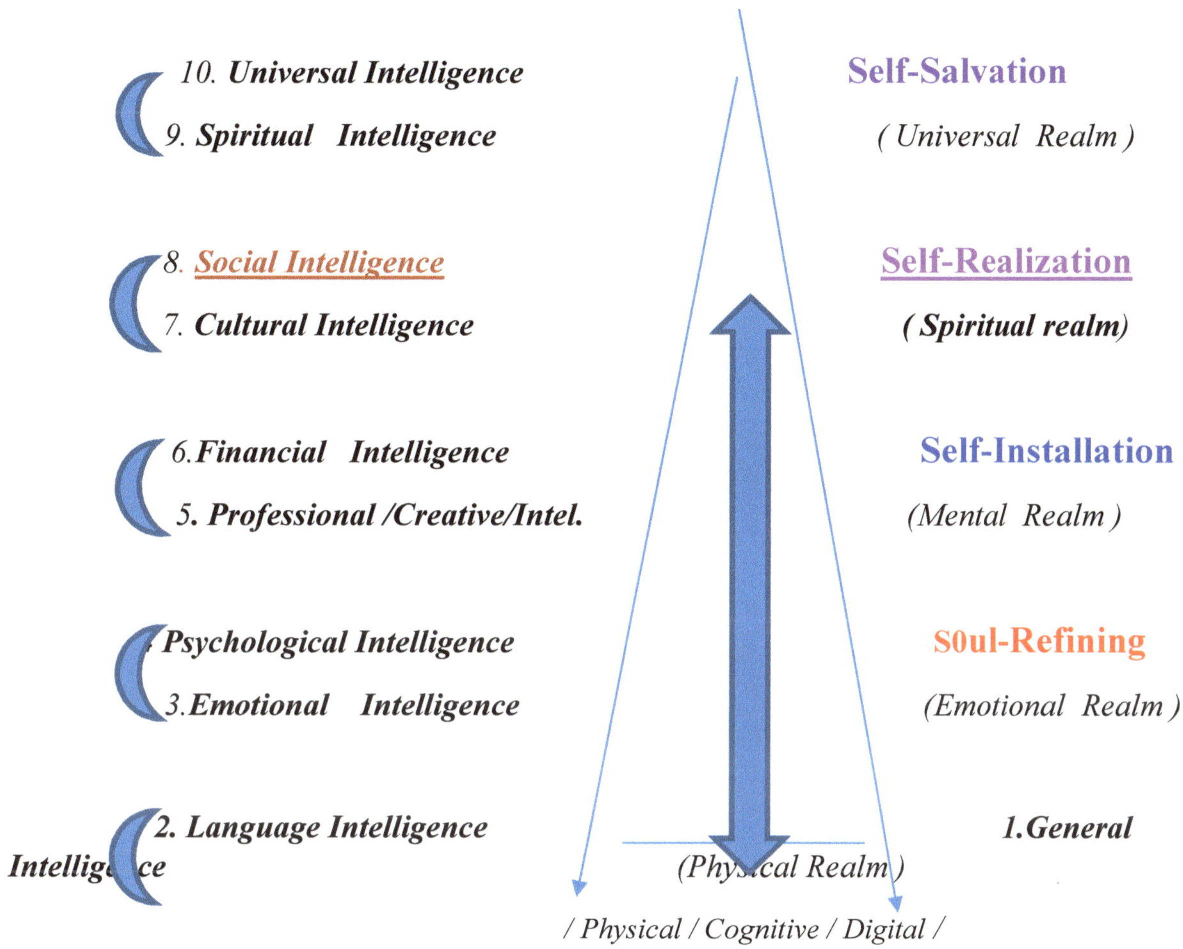

Synthesis-Analysis -Synthesis

Inflate Your Personal Pride with Social Might!

Generalize - Internalize - Personalize - Strategize – Actualize! Socialize!

Be Wise!

Don't be Socially Negligent. Be Socially Intelligent!

1. Social Awareness and Human Aptness

Social Intelligence is the path of our obtaining global knowledge in cosmology, genetics etc.in life is an integral process of a gradual self-transformation in five dimensions, a considerable enrichment of intelligence and self-awareness becomes a reality. ***"Humans are inherently social beings; that's our evolutionary legacy***." The quality of our lives now is the quality of our **SOCIAL INTELLIGENCE or** our global, digitally enhanced vision of the world.

"Anti-social behavior is a trait of intelligence in the world full of conformists."

Evolution through relationships shapes us personally and socially. No doubt, socialization promotes our intellectual culture. However, relationships have never been so problematic and conflict- bound as they are now.

"***Our collective reality is a symbolic expression of fear and an insatiable desire for security, as well as heavy layers of negativity and judgmental attitude that we have accumulated in our collective human psyche***." (*"Scientific American Mind," (July 2014)* **"A shift in planetary consciousness is taking place now."**

As with any other intelligence, ***social intelligence also needs to be holistically developed*** because it incorporates all the vistas of intelligence outlined above. An unenlightened person usually has a narrow vision of himself and life as if he were looking at the world through a small window. However, when a person expands his self-consciousness, it's like going through a door into a world of communication. Every one of us is globalizing his mind, vision, his / her personality, and professional outlook. "***Instead of fighting the darkness, bring in light. Allow all people to be as they are. Shine for others, too!" (****Dale Carnegie)*

Being ***socially apt*** means **establishing a mind-to-mind and heart-to-heart connection** with all the people you contact within face-to-face digital communication. If we remain impersonal, we cannot have a peaceful, healthy society. Learning to expand our living intelligence holistically, we are also developing new respect for life in yourself and humanoids because our neurological circuits are implanted into them.

The vision of ***a person or a humanoid*** who **has obtained Living Intelligence in five levels,** becomes inwardly stronger. The mind performs correctly and in unison with its trained emotions. People gravitate to him / her for spiritual energy and life wisdom.

At this spiritual level of self-training, you are also obtaining the **Ultimate Result Vision** of yourself. (*URV*). You are changing the ***form and content*** of your life; you are becoming ***the Vintage you.***

Having read the book that far, I can vouch that you have developed a solid trust in yourself, and you can manage any situation consciously. You are at the point of self-revelation and self-realization! Congratulations! You can rightfully declare,

I Am Free to Be the Best of Me!

2. Social Metabolism without Any" Ism"

Forming the ***holistic content of your life and*** propelling yourself forward with ***a much more refined self-concept, you*** become socially apt, more responsive, and responsible, and that is a great accomplishment of our unprecedented socialization. Changing ourselves, we create a new ***holistic WE*-CONCEPT** of social unification and mutual responsibility for life on planet Earth.

Our new digitally generated social connection is gradually forming a new world community of* GLOBAL HUMAN CONSCIOUSNESS *that has not changed for centuries.

Information is ammunition because our awareness of it changes the content of our lives!

Unfortunately, due to high tech explosion, life has a different narrative of self-erosion! W***e live in a socially*** and digitally polluted world. the world of social labels with ***the "ism" endings*** of all kinds. It has become an urgent necessity to disintegrate ourselves from the social pollution and formulate our own personal standpoints, finally ending our dependence on *"**the collective unconscious**." (Carl Yung)*

Our **SOCIAL METABOLISM** is now clogged more than ever with ***impersonal egoism, nationalism, chauvinism, and racism***. Many people do not think on their own; their judgement is formed by *the* ***crowd mentality and the mass media indoctrination. Mass media does not provide a holistic, objective, sincere view of the circumstances that generate the events.***

The Law of Cause and Effect is not observed by our media!

Being in the dark, people often adapt to the commonly programmed view. Total unawareness, wound up by the manipulators of the minds, generates impulsive reactions of the *"**collective unconscious," per***ceived as the fight for justice that, in fact, is the struggle against a holistic, insightful vision of life. ***Each of us needs to learn to think on his /her own, based on well-processed social information and his / her own social metabolism***

***"Our seeing of the reality requires a correction of mind; just as clear vision requires a correction of the eyes**" (Alan W. Watts)*

"Autonomy of thinking is a Must!" (*Reid Hoffman)*

The content of each individual life needs to be sifted off its polluting stuff! Negative people, like redundant information, need to be sorted out by you, too. Our interdependence on tools, manipulating social opinion, entangles the. minds and blurs ***the objective social vision***. So, let's be governed by ***aware attention*** and purposeful ***sorting out of the social contacts*** that we have. We have no luxury of wasting time on the people of no intelligence and no human value, killing their lives in meaningless talking and crowd driven actions.

Socializing means Getting Interested and Becoming More Interesting Yourself!

3. Mind-to-Mind + Heart-to-Heart is a New Social Art!

There is the reason I place ***Social Intelligence*** at the top of ***the Self-Acculturation pyramid.*** To learn social skills, a person needs to incorporate all the above studied intelligences and become a new human being - a really advanced one, a man / a woman of integrity, able to combine thought and feeling and be socially fit. A higher level of interpersonal mental frequencies is needed. ***A new level of social vibrations is changing the world.*** And now, we need to come to terms with the human-robots relationship forms!

Our Mind-to-Mind with AIs should become a Digitally Enhanced Art!

Any art is a form of communication, but this form becomes classic if a piece of art enclosed in it radiates beauty, has a structured composition, and is very thought-provoking. I think that our socializing in general and the one on the Internet should be much better both ***in form and in content*** and our Internet companies should contribute more to the quality of our socialization.

The vital principle of holism is ruling the world of unification for human salvation!

Amazingly, socialism was uniting people, while capitalism is dividing us. Obviously, the best features of both social forms are to be adopted by us on the new twist of the evolutionary spiral. The countries that have incorporated ***the features of both social systems*** into their national framework are doing much better than the rest of the disconnected world.

Sweden and China are the best examples of such conversion that results in a much better standards of living.

The concern for the well-being of the people needs to be driving us toward unification, and that's the only way for all humanity to survive. Social behavior has evolved unbelievably because of modern communication technologies. With the younger generation leading the way, large groups empowered with Internet-enhanced communication accomplish a variety of diverse goals.

We are gradually becoming global software entrepreneurs, and we develop electronic ***competences of holistic value every day***. **We enhance our souls and expand social outreach with a digital switch!** However, ***our social communication should never lose its individual characteristics*** *(See the Chunk 1 above).*

Often, inept socially people do not integrate collectively and are believed to be anti-social .

But, in fact, the less we socialize , the better we are!

However, these people are better thinkers, and they can contribute more to society than the ones that merge completely with the massive culture and lose their individuality. In fact, we get back here to ***the first level of intelligence – the physical level.*** It's impossible to help and love others, without first doing your own self-quest and giving your best to society.

The Identity of the Hearts and Minds Binds!

4. Social Network Equilibrium

The difficulty some people have in global communication comes from their ***different mental wavelengths*** which do not coincide with the vibratory frequencies of the people they communicate with. . *"**It is as if we tune in to a certain broadcasting station on a wrong frequency**."(John Baines)* A mind-to-mind communication becomes possible only between the people of adequate intellectual frequencies that decisively influence ***the quality of*** **CONSCIOUS COMMUNICATION** between people. Intelligence removes vibratory disharmony and establishes a social equilibrium of mind-to-mind contact of mutual understanding without being politically bereft with a war-mongering mentality.

Such intellectual chemistry has much greater social potential!

New social skills that we need to develop at this level should not conflict with a sincere intention to contribute to the world our unique gifts of intelligence. With the unprecedented speed of life that stresses us out unbelievably, we need to establish mind-to mind and heart-to heart connections between people that we communicate with for business' sake or personal reasons. Our new social experience proves that aware attention displayed on the mental level and emotional understanding, backed up by **cultural intelligence,** ensure much success in business solutions and ensure that a person will not get lost in the ocean of technological havoc.

The holistic paradigm *(Synthesis-Analysis -Synthesis)* ***needs to be followed religiously in five dimensions.***

Generalizing- Internalizing- Personalizing -Strategizing-Actualizing!

The contribution of our computer mega companies, like *Amazon, Face Book and YouTube, LinkedIn* is huge , and the improvements they are implementing on this path are most promising. Social pollution puts us in front of the necessity ***to X-Ray social contacts*** for their personal, emotional, mental, and spiritual validity for us

People that we allow into our lives shape our lives in every dimension! The question is how!
The dynamics of socialization at present are extremely and ruthlessly competitive, but our social metabolism should grow intellectually and in a holistic way.

"***What makes the meaning of life is people. So, try to be good to people immediately around you and in a broader, world community."*** *(Reid Hoffman)*

However, if you observe unconscious behavior in your partner, hold it in the loving embrace of your knowing, **so you won't react**. You need to respond consciously and intelligently and cut the person off your social loop at once. Intelligent networking is extremely beneficial for self-creation; learn to establish it. *The Scientific American Mind," Nov. / Dec. 2014 writes,*

"A single brain is a computing marvel, but it pales in comparison to the powers of the net-worked mind."

5. Social Connection is Language Interwoven!

For our digital communication to be more harmonized, we need to become ***more polished in our language competences***. The correct use of the language, the tone of the language used, and the vocabulary chosen to get the needed thoughts across are crucial now. *(Check out Language Intelligence above)* It's good that the use of profanity is being checked by face book, but we need such attention to be paid to the quality of the language we operate with on every sight. A human being at the other end of the virtual line is as real and perceptive as you are. Both verbal and digital communication demand your **thinking first then saying something or writing a response.** We should not generate negative consequences for ourselves or other people due to our texting on impulse and transmitting this impulse to millions of others. "

Remember," *90% of people do not care about your problems, and the other 10% are glad you have them*" *(Carl Yung*) A great psychologist means that we need to be centered on confronting our own problems ourselves, "***being the thing in itself***" in solving them because being part of all life, you are still solely responsible for your own life and its outcome. ***"You cannot conquer life because the part cannot conquer the whole, but you can be the best part of the whole!"*** *(Carl Yung*) "Disconnection is death!

The new Self-Consciousness formation is Our Salvation!

Being a part of the whole is also speaking the language of the whole, called "Globish. " English needs to be respected for its ***unifying quality*** and the incredible ***feeling of belonging and being One in the context of life*** that it has granted to us with its mathematically- impeccable structure that , of all the world languages, allows the machine to program the information in the best way possible. *(For more on the universality of English, see the book " Language Intelligence or Universal English!"*

The information being programmed in English is the greatest accomplishment of the Internet!' *The dawn of human brain-to-brain connection has arrived with English***"** *(Rajesh P.N. Rao) So,* ***let's not distort this unique language while texting , speaking. or writing.*** To grasp the complex problems of socialization with its political conflicts and war-inflicted sorrows, we must bring about the understanding of ourselves and others in synch based on emotional and psychological intelligence *(See above)* in accordance with the **LAW OF THE_RIGHT HUMAN BEHAVIOR.** (*R. Wetherill/" Right is Might!")*

That should be the core of the **GLOBAL INTELLIGENCE** that we are developing slowly ,but surely. Obviously, we need to focus now on working out ***The Code of the Global Culture*** backed up, possibly, by the United Nation's Charter of International Communication. Only then will we be equipped to master our national and political differences and focus on raising **GLOBAL SELF-CONSCIOUSNESS** ***that*** will propel us to a new level of extra-terrestrial communication and be accepted by the **Universal Community of celestial infinity**.

Self-Transformation benefits the World at Large very much!

6. Synergize the Trajectory of Your Life to Socially Thrive!

Electronic globalization promotes business, scientific discussions, and all kinds of networking that makes our overall intelligence more creative and much more beneficial." Globally ***competitive professionals work within strong networks, achieve differentiation in the society by developing their network intelligence***. " *(Reid Hoffman)* As technologies improve, people are finding more creative ways ***to utilize their social skills*** to their advantage, backed up by major social sites, such as Facebook, Tweeter, LinkedIn, and others

. **We are becoming increasingly socially apt in our social behaviors.** I have noted above how important an individual ***value preservation and correct language use*** are in this process, and how vigilant we must be not to be swept away ***by a common trend of imitating others*** in the search for immediate gratification of our needs. With robotics developing rapidly now, our social hunger will be satisfied even better. ***Robots will soon become our companions***, more understanding and perceptive than humans. In recent years, a growing number of researchers has demonstrated that making robots more social "***is creating the perception that they are more sentient than a mere machine.*** They will make us more sociable and even more global. **They will expand the network of our intelligence.**

Meanwhile, ***becoming global*** is "*growing among those who know career rules and have the new skills of a global economy ("The Start-Up of You").* It translates into the necessity to assess the basic competences. An individual should comprise ***an individual plan of running his life*** as his / her personal business, "*investing in themselves, in their network, and in the society."(R. Hoffman)*

We can master basic intelligence better if we are given digital tools for life learning.

By reaching out to a network of knowledgeable, hardworking people who display professional awareness and social competence. W

"We can bring more depth and understanding to our life goals , gaining worldwide interest and support at the speed of the social Web*"("Forbes," The World Almanac of Wealth< 2014)*

Another crucial aspect for the present-day economy is **the ability to promote science and the scientific thinking** that in turn "***creates new concepts, new ways of understanding the world that have an enormous influence on how we think and act'*** *(" WIRED," 2012, Barky Schwartz)*

Mr. Schwartz means that when we think about **the technological impact of science**, we tend to think of the things that science has produced, but there is another kind of technology produced by science thanks to the technological revolution that has just as big an effect on us as technology. A man of a great scientific charisma, ***Dr. Neil deGrasse Tyson*** reminds us,

"Scientific Literacy is a Must*!"*

"We Might Call It Idea-Technology."

(End of Stage Eight- Social Intelligence)

Stage Nine

Super Level - ***Spiritual Dimension*** – ***Self- Realization***

Spiritual

Intelligence

"To be spiritual in truth means only one thing that spirit is manifested in one's own brain."*(John Baines)*

Put the Shield of Intellectualized Faith in Front of Your Face! To Believe Means to Know!

God is Omnipresent. He is in You, Me, and the President!

An ancient medallion with Prophet Moses and ten Commandments that he brought to the world.

"I Am God, and I Am One!"

The Intellectual Odyssey of a Seeking Soul

Going beyond the Terrestrial Boundaries

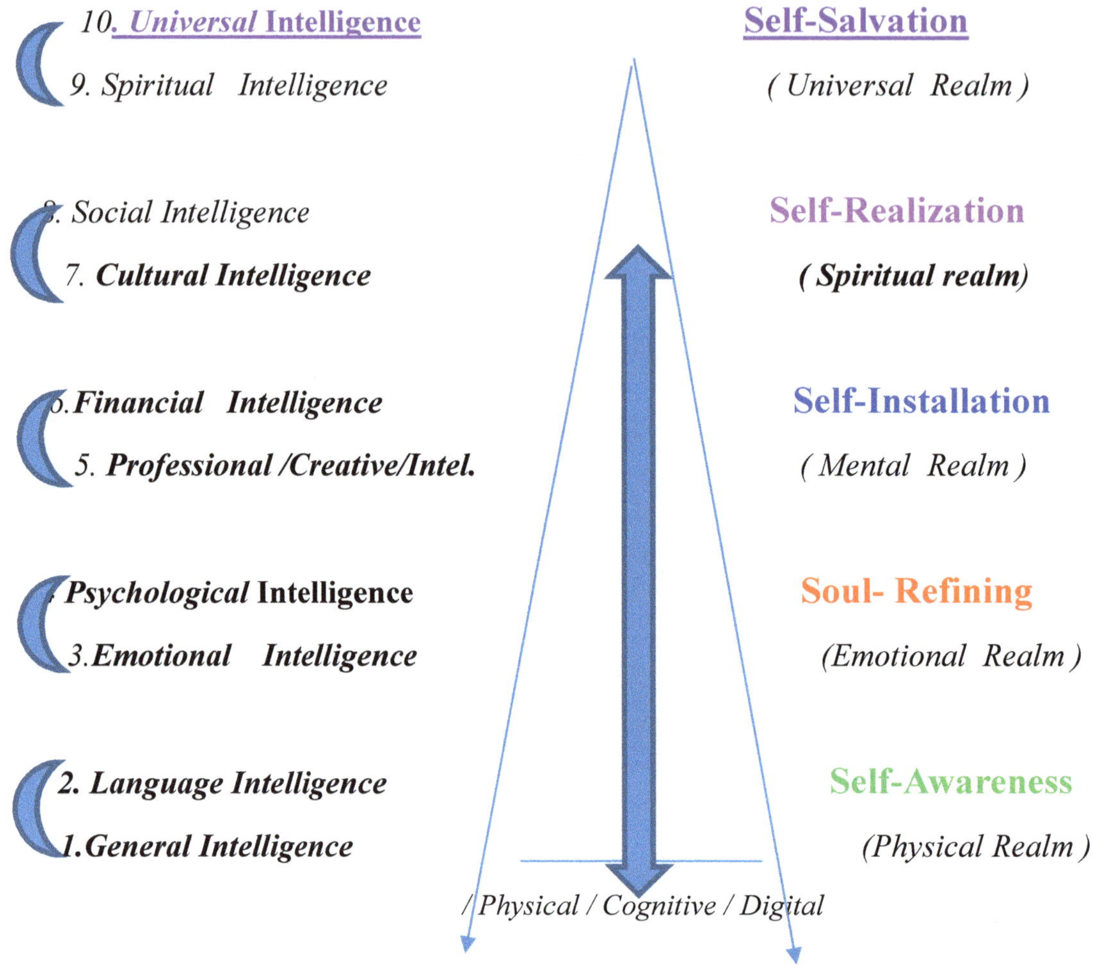

The Fractal of Our Spiritually Intellectualized Self-Growth:

(Body+ Spirit+ Mind) + (Self-Consciousness + Universal Consciousness)

(The physical form) + ***(the spiritual content of life)*** =

Self-Awareness + Soul-Refining + Self-Installation + Self-Realization + Self-Salvation!

A New, Whole, Spiritually Intellectualized, Wise ,and Self-Emancipated You!

1. We are All in the Court of Almighty God!

I try to live spiritually

And not process my problems virtually,

For as it is Above,

So, it is below!

I am strong in my spiritual reactions,

But I am weak in my physical actions.

So, for my New Year relation,

I made a new declaration.

If something is wrong in my life's pact,

I don't declare it as a fact!

I ignore it and disregard,

but I am fully aware of that!

I try to declare

"I am OK here and there!!

And to the question, "Are you happy, or what?"

I answer with a sure nod.

I am happy I am alive,

Healthy, beautiful, and in love.

So, I thank God every day and pray

For being blessed with a new day!

"Life at Its Best is the Life of Faith!" *(Dr. Stanley)*

2. The Path to Spiritualized Intelligence

Transcendent You is a God-Centered You!

(See the book "Transcendent Us and AIs"/ Universal level/ Digital psychology))

There is no personal, or moral growth without **SPIRITUAL INTELLIGENCE.** ***Spiritual -Moral maturation*** is a lengthy process that needs to start with a willful ending of judging others and starting ***to be more objective*** about one's **Self-Acculturation**. ***"To harvest good fruit, cultivate good thought about yourself and others***." (*Buddha)* Our values, beliefs, behaviors are over-disposed by our manners, hectic living habits, our being raised in a certain faith, our being conditioned by the mass media and society at large. In his mind-changing and soul-forming book "*Right is Might*," *Richard W. Wetherill* writes:

: "Moral relativism (*no right / wrong ethics*) is ruining our souls."

We must halt any images of immorality and fear that harm our souls on the path of self- bettering. It's paramount ***to see an image of a positive moral outcome*** in any challenging situation through the mind of a growing child. Just Sunday visits to the church will not work. .Many people piously light the candles in the church, but ***they blow off their inner light or the light of their loved ones outside the church***. Be sur you have the five most important aspects of life in synch and preserve your ***moral integrity.*** *Also, stop justifying yourself for any harm done to anyone.*

Sacredness and Nobleness — *Universal*
Morality/ Faith — *Spiritual*
Intelligence — *Mental*
Attitude — *Emotional*
Health — *Physical*

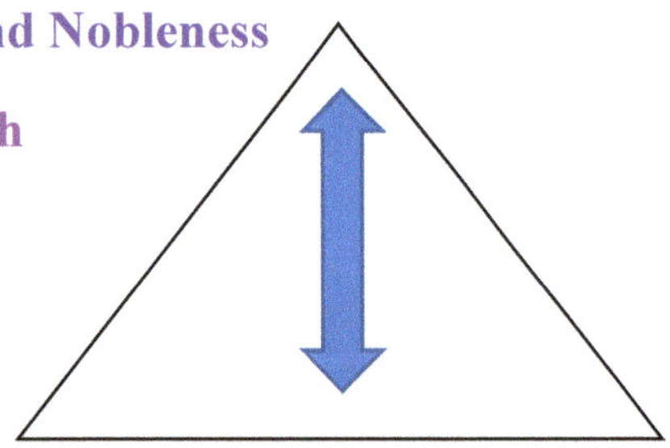

A moral person is supervised by his / her intellectual spirituality because there is no spirituality without true, heart and mind processed intellectuality. No wonder, the spiritual dimension of self-growth goes after the mental one. So, when people complain that God does not hear their prayers, that's quite explainable. There is no true intelligence and wisdom in a person. He / she is unable to ***establish a common spiritual line with God*** that is, in fact, ***Universal Intelligence*** or ***Super Consciousness*** that we can tune to only if our self-consciousness is high enough.

(Body+ Spirit+ Mind) + (Self-Consciousness + Universal Consciousness)

Self-Worth is a Multi-Dimensional Course!

Nobleness + Morality + Faith = Personal Integrity in Place!

3. Bad Morality Does Not Give Life the Chance to Happen!

A morally uncontrolled person is living under the ruinous effect of ***progressive demagnetization and de-humanization.*** It's a downfall at every level , and it inevitably brings him / her to the point of no return – breaking of the trust , cheating, divorce., guilt trip, depression, etc.

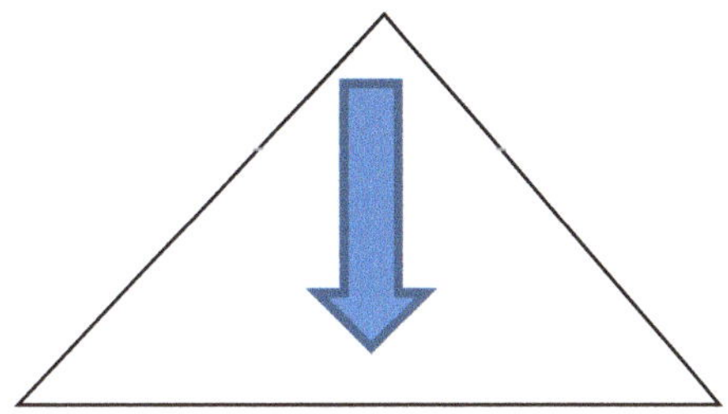

Lack of Morality / Lack of Faith	***Super Level***
Lack of Intelligence / Ignorance	*Macro Level*
Laziness / Inertia / Lying	*Mezzo Level*
Negligence / Automatism	*Meta Level*
Impulsivity /Lack of Responsibility	*Mini Level*

Neither of us is perfect, but it's especially important to realize our deficiencies and **correct our personal defects consciously.** The simple and the most basic qualities above will serve as a point of reference for you.

See if you need to work on any of these areas t***o become a spiritual aspirant*** and start creating yourself in this direction **holistically.** Otherwise, the process of moral demagnetization will start snowballing.

Moral demagnetization is the beginning of de-personalization, too.

To stop the downfall of personality development is not altogether impossible, but it requires a lot of self-monitoring and self-control. **SELF-MONITORING** cultivates an inner balance of ***morality and personal integrity*** that become the basis for a personal quest and INNER **SELF-GRAVITY** of a truly nobleperson ***with a heart and mind in sync.***

Intellectualize the heart and emotionalize the mind. Be One of a kind!

Present-day science tries exploring the concept of impulsivity that breaks us in half and pushes us away from balance. Rationalization of thinking is essential in this respect.

To accomplish that, we must halt the unconscious, automatic way we think and justify any whim to please oneself without delay! In his groundbreaking book "***Robots' Rebellion***," *David Ickes* writes, " ***The further you stray from balance; the more extreme life becomes***. " *David Icke* states further that if we go too far to the positive polarity, we lose touch with practical side of life. ***"Positive energy needs a negative balancer. Only balancing yourself between the two polarities of life can link you to the highest levels of consciousness.***

Balancing oneself morally is the skill of the inner equilibrium feel! *(See the book "Soul-Symmetry," Canada, 2021/ Video on You Tube)*

Your Moral Make-up Must Be Under the Thinking Cap!

4. Being Spiritually Aristocratic

Spiritual Intelligence is based on faith in God and yourself, or in Super ***Consciousness*** and your own ability to be always connected to it through your ***Self-Consciousness.***

(Body+ Spirit+ Mind) + (Self-Consciousness + Universal Consciousness) - *A Connected You.*

Dis-connection is Death!

Our spiritual essence is another factor that will never be simulated in humanoids. We can program their neural system for noble emotions to prevail in them if our own inner system is based on unshakable faith that always characterizes ***the men of their word and the people of action.***

There is no Self-Salvation without sincere faith and Self-Transformation!

One episode in the Bible is very illustrative of our shaky faith, When Jesus Christ urged Peter to walk on the water with Him, Peter started sinking " ***You are a man of little faith. Look at Me, not down into the water."*** Many people give up on their self-growth due to the lack of faith in them, Self-Resurrection is interwoven with faith, self-worth, and character.

Most people just walking in the street, look down at the ground, not in front of themselves for the whole picture of life to be spread before your eyes The Sun in the sky remind you of the eye of your spiritual leader, whoever it might be. All the true spiritual leaders' prophets, disciples, gurus , etc. are always the men of their word, character , and action.

They inspire us to stay in the light because they have a sincere ability to be the **LUMINARIES.** They inspire us because they are inspiring themselves with the spiritual intelligence that they have accumulated.

Thus, the top of the intellectual development of an individua is his / her ***spiritual intelligence that is universally bound***! It means a new vision of the world that **bridges science and religion** and turns both into true beyond ***the terrestrial spirituality*** that liberates the spirit of a believer, uplifting him / her to the new heights of human evolution.

SPIRITUAL LIBERTY DEFINES OUR EVOLVING SINGULARITY!

It's the process of a ***serious self-quest and self-perfection,*** not self-demonstration, and self-confirmation in a religious sense! Being spiritual also means that you have matured in **self-**awareness and life-awareness **physically**, leant to monitor your life **emotionally**, self-installed **professionally**. constantly tame yourself **spiritually** and feel holistically united to life around you **universally**. The stages of self-growth are all in a holistic unity of constant, digitally monitored ,but self-mentored and self-monitored **SELF-ACCULTURATION.**

Self-Awareness + Self-Monitoring + Self-Installation + Self-Realization + Self-Salvation!

5.Training Humanoids Should Also be Five Directional.

With such an inner current, you will be shaping yourself into an ***accomplished human being*** with a new perception of life - new **thinking, speaking, feeling, preserving, and acting**. You are in control of your intelligence, not just your society-shaped intellect.

Your WHAT + WHY + HOW are personalized and in synch, and it is based on the knowledge that you had accumulated, selected, and actualized.

This process ***refines the soul,*** and it starts shining from inside, irrespective of the skin color. I have met many people of color in the USA that are shining from inside with true readiness to help and to back me up with a blessing at the moment that I needed most, being often lost in the country with a hostile, capitalist structure , set up on the money-making and an impersonal attitude to people.

Spiritual immunity and liberty build up our global intelligence and religious unity!

Many people in different social offices have a nice smile on their face, but they don't care about what bothers you. They go to the church every Sunday, though, believing to be Godly, but being, in fact, heartless and godless. So, being **SPIRITUALLY ARISTOCRATIC** is a challenging responsibility without a devilish inner response, "What **do I care?** or **What is there in it for me?** your life will have a renewed quality, depending on the state of your **SELF-CONSCIOUSNESS** that we need to raise in our kids and ourselves as the main goal of life.

So, ***our faith must be coordinated with the newest developments in science*** and the most advanced AI applications, intellectually challenging and respectful of human intelligence that needs to work, not just utilized for the needs of an intellectually lazy mind. ***Also, such collaboration must be of a much higher mind + heart ethical ration, leading us to***

TRANS-HUMAN SELF-ACCULTURTION.

(See the book "Transhuman Acculturation"/ Spiritual Realm / Digital psychology for Self-Ecology

It means sincere, consistent, and conscious ***culturing of new beliefs, thoughts, words, feelings, actions, and the entire destiny through digital means.*** The possibilities are breath-taking, and with *Elon Musk's* sense of purpose, we are learning to think beyond planet Earth.

Digital Self-Acculturation is sculpturing a human being in you as a Self-Monitored guru!

The urge to *express* ***personal exceptionality*** *and* ***human superiority*** will not let *Artificial Intelligence* enslave us. Our ***unique human individuality***, ***creativity, and incredible ingenuity are*** the basis for our life infinity.

Human ⟹ Trans-Human ⟹ Super-Human!

Sculpture Yourself for Transhuman Might!

6. Practice What You Preach!

To conclude, my overview of this important stage of acquiring Physiological Intelligence, let's consider some more suggestions and the necessity to do self- X-raying that help ***synchronize our emotional and psychological make-up.*** . The first wonderful piece of advice is given to us by Aristotle in "*The Nicoma Bean Ethics:*"

1." ***Anyone can become angry-that is easy. But to be angry with the right person, to the right degree, at the right time, for the right purpose, and in the right way – this is not easy."***

2. **Synergize Your Space!** Our immune systems are very vulnerable now with the electronic pollution mounting up and engaging us totally mentally and emotionally. Only your positive reinforcement or your self-illumination can do the job. So, ***bank on your personal synergy****!* *Synergizing is much better than compromising that we often must resort to, anyway.*

3 "Perception is more important than intent!" This rule is crucial for our emotional and psychological health. Whatever a person of personal contact is, his perception of you, the space around you, your voice, your words are much more important than the initial intention you had

You might have worked it out in a wrong way, and the person perceived it wrongly. But it was your fault! You're trying to justify yourself with, **"But I didn't mean it!" It doesn't work here**. That was the way you had communicated your thoughts, and that was the way it was perceived.

So, **say what you mean and mean what you say!** *("Language Intelligence or Universal English")*

4. Always neutralize an opponent's intention for control. In every encounter, one person has more command over the situation than the other. If you happen, be the one, be diplomatic.

5. Build a shorthand profile of the person facing you. X-Ray your conversation partner for his/ her weaknesses and strengths. You'll manage the situation better mentally and emotionally.

6. Trust your intuition. It will scan anyone without fail and help you be more diplomatic in a mind-to- mind battle. ***So, dress up your face and open your heart !***

Be Exceptional at that!

7. **Practice language mindfulness, that is, moment-to –moment awareness of what you say**. Focus on the structure of your thoughts: the point, the proof, and the bottom line. **(General-Specific- General***)* *(Cognitive Intelligence, Chunks 7,8)*

8. Be focused on what other people say, hear them out. A dissipated mind is a constant stress and an irreparable breakdown of the psyche, both yours and your partner's. **Listen to hear!**

"Dissipated Consciousness is a Lost Life!" *(Carl Yung)*

Smile to Synergize Your Space and the Eyes!

7. Be Kind to the Unkind. Be One of a Kind!

9. Practice paying aware attention to the logic of your communication. Be able to establish a mind-to-mind connection. Get rid of redundancy in your speaking. ***Don't verbalize everything on your mind. Sort your thoughts out.***

10. Continue practicing the use of self-inductions in any situation. Give yourself a boost of confidence, self-love, self-defense, and positive self-attune. ***Be consciously alert!***

11. " ***Come from the divine source of love.*** Generate love in your heart and mind!" *(Dr. Dyer)*

12. ***Clean your mental-emotional landscape!*** Let constructive boosters root in your mind; let them help you weed out all the doubts, bad memories, and emotional inadequacies.

13. Think and act in a win-win mode. Do not generate negative energy - synergize!

14. Don't criticize – help! **Don't politicize – help! .** ***Don't compete – help!***

15. Don't second guess – talk it out! Less talking = less pressure and more result!

16. ***Don't judge – help!*** We are all much more alike than different. Be in a hurry to understand.

17. ***Don't react – respond consciously!*** People always have something to be respected for. People are innately good and what lies within is even better.

18. ***Be just divinely guided, looking for the way, but do not block the sun for others.***

19. Be kind to the unkind. Be One of a Kind!

Doing all that, you'll increase your own awareness. You'll also affirm yourself inwardly and outwardly and give others psychological air. Giving others psychological air, you'll create an environment for synergy.

This book is not meant to be read from cover to cover. It is a study book and **a help-help one**. That's why I try to make every chapter a page too long, follow the holistic paradigm- **synthesis-analysis -synthesis.**

20. Remember the main mind-set that I keep posting **in all five books,** (*See Book Structure*) promoting the ***Inspirational Psychology of Self-Ecology:***

I AM MY BEST FRIEND. I AM MY BEGINNING AND MY END!

Please note please, you are not supposed to read the books consequentially. Pick the realm of life (***physical, emotional, mental, spiritual, or universal)*** that you need to fix most and go from there.

"Life is Not Happening to Us. It is Responding to Us!"

(Sadhguru)

8. Delete Any Prophecy of Your Spiritual Odyssey!

Finally, your **SPIRITUAL ODYSSEY** is a life-long journey that should start with the kids. Their spiritual values are extremely poor if at all. Many of my students complain that they can take a child to church only if ***a visit to MC Donalds*** is promised after that.

The kids of the immigrants are impacted by their **ungodly Americanization and its "civilized barbarism**" *(Carl Yung)* the most. While this country is shaping us into ***the international mob*** and making us all more characterful and purposeful, ***we need to preserve our national uniqueness in our kids***.

Our parents are our tests for spirituality and inner sacredness.

The rule below is the main one the immigrants live by here, but it involves the **SPIRITUAL IMMUNITY,** too presupposes raising of **SELF-CONSCIOUSNESS** in any reality. So, keep your life away from any fraction. Oversee every thought, word, and action! The cultural depth of every country should be preserved, and its uniqueness respected everywhere de facto, not just de juror. We declare it, but we do not feel it**.** ***We are not consciously aware of*** it.

Our self-consciousness must be intellectually spiritualized!

Transforming yourself intellectually, you develop your spiritually, developing a new outlook-new thoughts, feelings, convictions, values, and aspirations. ***You are different because you embody different, refined patterns of behavior, controlled by your conscious mind.***

You are not part of the "COLLECTIVE UNCONSCIOUS. *"(Carl Yung)*

You respect yourself, and you feel content with what you are. You generate new abilities to love, to forgive and to forget because you have sorted them out through your **growing spiritualized intelligence and self-consciousness.** You have a strong position in life, and you can stand up for yourself, and the others, weaker ones. ***" I use my own brain, and I never stop borrowing the brains of the others."****(A. Einstein)*By

There are many scientifically minded **GOD-TRANSMITERS** now, and they are amazingly insightful and self-educated. They gather thousands of people that tune to them physically, emotionally, mentally, and spiritually because their mental and spiritual frequencies ***synchronize with the God's, or the Universal Intelligence messages.***

The crowds that *Joe Osteen* gathers are all there to get inspired by intelligent minds and a pure soul. ***The sincerity and the objectivity of the knowledge presented do the trick*! "I want to know the God's thoughts. The rest is details**!" *(Albert Einstein)*

Obviously, just praying is essential only when it is *consciously mind-changed and in no way mechanical.* When we tune ourselves to the station **"God, we *excavate the best qualities in us."***

In a Noble Life Quest, I Am My Very Best!

9. Being Godly in a Godless World

Life is the process of self-seeking, and many people are on this journey now. ***We need to see the religious values through the prism of new science accomplishment, dis-entangling ourselves*** from the old dogmas and the limited scientific outlook

I have been outlining the nine levels of intelligence and their role in self-monitoring so far only to remind you in each chunk of information that we are evolving our intelligence to enrich our spirituality and to raise our self-consciousness. It should be done**,** not just in a religiously mechanical way, but through deep, insightful, aware, unshakable awareness that we are being digitally governed from the Above, or, actually, everywhere, by some creative force:

Universal Intelligence, the Master, Mind, or the Source that we all perceive as God and call God. "Man may produce the stumbling stones; God alone-in the mind of man may make them steppingstones." *(Edgar Cayce*)

I think that every sacred book I HAVE read is , actually unfolding the secret of life in the five levels: ***physica**l (mini); **emotional** (meta); **mental** (mezzo*); ***spiritual***(*macro*). ;and *universal (super). Thus, we* are developing ***the Super Conscious Mind***

I'm No Longer Automatic; I'm Becoming Spiritually Aristocratic!

Mini level - **I personalize myself mentally, physically, and verbally.**

Meta level - **I emotionalize my mind and get more psychologically– aware.**

Mezzo level **I intellectualize and individualize myself professionally!**

Macro level – **I acculturate, socialize, and spiritualize my life.**

Super level **I perceive with my soul; and I see myself as part of my whole life.**

The body is ***a physical, emotional, mental, spiritual, and universal part of the entire life*** in the Cosmos ,and therefore, it needs to be sculptured in all these planes Self- Evolution is a gradual process, and it is going on according to the Universal Plan. Each level is integrated with the next one. Self-realization is the process of internalizing the necessary ***intelligence*** in synch with the development of ***self-consciousness.***

We call this union - ***personal integrity***, and it manifests the outcome of the self-actualization process that ends with life itself. So, when talking about ***Spiritual Intelligence, I*** do not mean any religious affiliation, nor do I give preference to any religion. All religions are the interpretations of the eternal concept of creation **- God.** As ***John Baines*** puts it in his wonderful book (**"*Stellar Man.,)***

"Spiritual Aristocracy begins with the Individual and Ends with Him."

10. Every Soul Lives on Universal Nutrition!

RELIGIOUS WIDOM ***that humanity has accumulated for centuries instills in us the basic universally acknowledged values and connects us in faith.***

We stop discriminating against the believers that are different from us and start **becoming GODLY** and considerate of our common humanness that is now instilled in a machine whose national belonging does not matter. Humanoids may be aggressive in their general perception that is *not yet neurologically regulated*, Spiritual exploration is the matter of your Intelligent Consideration!

To evolve, we all need spiritualized intelligence without negligence!

Acquiring spiritualized intelligence is our common goal that cannot be achieved without our new "**SPIRITUAL LITERACY"** that *Dr. Neil de Grass Tyson*, our leading cosmologist, is calling on us to have as "*science literacy*" .But science and religion are being united now with our **QUANTUM MATURATION**.

It will help us intellectualize our hearts and emotionalize our minds, or put our hearts and minds in sync that, in my understanding, true spirituality is.

But there is no wisdom, without the wrapping of a man's intelligence**.**

"Spiritual intelligence uplifts and expands our souls, brightens our soul light," *(Fred Bell)*

The soul's light protects us against the poisonous impact of society. It builds up the spiritual immunity against godlessness of the people around that manifests itself not in a person's different faith, but in the desert of his /her corroded , un – educated, dark soul. That's why before the spiritual dimension; there is the mental one**.** **To feel God, you must know God!**

There are a lot of different speculations made about spirituality in different religious affiliations, but it becomes increasingly evident with every passing year that all of us seeking one thing - ***spiritual maturity,*** *and it does not matter what religious way we choose to accomplish it. Being wise is also being in constant service to others and giving the best we have to the world. The Chinese wisdom rightfully teaches us that human values manifest themselves through service. Spirituality is the quality of higher intelligence, personal maturity, a lot of sacrifice, and our service to each other and God.* ***But "without sincere sacrifice, nothing is attained! "(****Buddha)*

However, we should not blindly follow the paths of any advanced minds, spiritual leaders , or a guru. You need to study, learn, process the exceptional, personal knowledge that you get ***through the personal grid of your perception*** and work out **YOUR OWN PLAN OF ACTION** and your own **MISSION STATEMENT of** the spiritual growth that is a life-long journey.

Conscious Religiousness + AI Upgraded Intelligence = Spiritualized Intelligence!

(End of Stage Nine - Spiritual Intelligence)

Stage Ten

Super-Level of Intelligence - the Final Step

Universal Intelligence

"It's time to make the voice of the Universal Consciousness heard."

(Sadhguru)

Our Infinity is in Universal Unanimity!

Don’t Be Spiritually Negligent. Be Spiritually Intelligent!

(Picture by Yolanta Lensky)

There is No Universality without Intellectuality!

The Intellectual Odyssey of a Seeking Soul

Going beyond the Terrestrial Boundaries

10. Universal Intelligence	**Self-Salvation**
9. Spiritual Intelligence	*(Universal Realm)*
8. Social Intelligence	**Self-Realization**
7. Cultural Intelligence	***(Spiritual realm)***
6. Financial Intelligence	**Self-Installation**
5. Professional /Creative/Intel.	*(Mental Realm)*
4. Psychological Intelligence	**Soul- Refining**
3. Emotional Intelligence	*(Emotional Realm)*
2. Language Intelligence	**Self-Awareness**
1. General Intelligence	*(Physical Realm)*

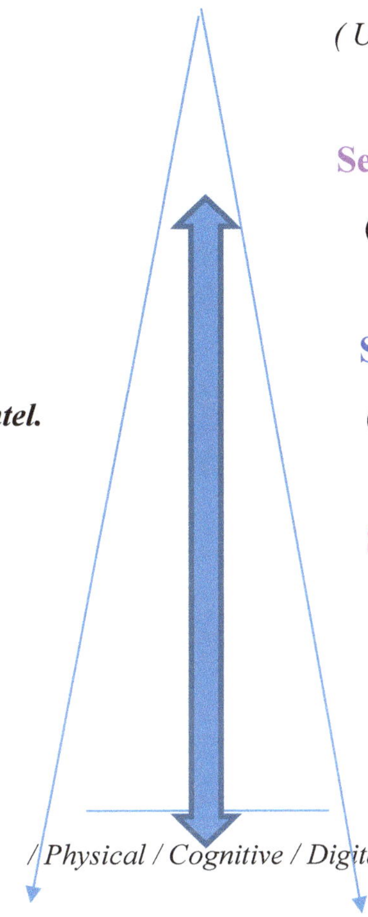

The Fractal of Our Spiritually Intellectualized Self-Growth:

(Body+ Spirit+ Mind) + (Self-Consciousness + Universal Consciousness)

(The physical form) + ***(the spiritual content of life)*** =

Self-Awareness + Soul-Refining + Self-Installation + Self-Realization + Self-Salvation!

A New, Whole, Spiritually Intellectualized, Wise, and Totally Self-Realized You!

1. We are One with Everything Under the Sun!

Attaining ***Universal Intelligence*** is forming a bridge with the Divine!

At this highest, super-level of self-creation, or the **SELF-SALVATION** level, ***"there is no separation between anything and anybody. There is only God, and everything is God!"****(Edgar Cayce*)

Elon Mask is a rare man who has obtained the universal mentality in his attempt to make a significant difference in our space life, and the scope of his ideas is absolutely genius! His words truly characterize people who have obtained universal intelligence. In one line with the exceptional people who have shaped the present-day world:

ALBERT EINSTEIN, NIKOLA TESLA, AND STEVE JOBS.

When we turn on the self-construction mechanism, we become extraordinary because we are governed by some higher forces, very much interested in our evolution. If we are persistent in our self-growth, we accomplish success, which is the most rewarding thing in the world.

At this level, we transform terrestrial knowledge into celestial wisdom!

Having formed at least a vague idea about all the above presented levels of intelligence and having worked on creating a better person of yourself, you have, for sure, gotten rid of much stagnant knowledge and beliefs, and you have stored some **LIVING INTELLIGENCE** in your brain and have sorted out the one you have accumulated so far. I am just up-dating it and systematizing some information holistically. **Awareness builds up confidence**.

Living Intelligence brings you in touch with Universal Intelligence in a new, much more reasoned-out and conscious way.

The profound questions about life in us and in the universe are not solved yet, but the fact that we are ruled by Universal Intelligence is not questionable anymore. Celestial wisdom gets transferred into the right action, and we become our own cheerleaders.

We learn***" to think, speak, and act right, and we refuse to think, speak, and act wrongly"***.(R. Wetherill) I hope that we will live much better and finally manage ***to orchestrate gladness*** in the heart every day. There will be less dissipated thinking and reckless, mindless actions on our part in the unpredictable future. that will also be characterized by the outcome below:

The willpower is firm, the values are moral, the actions are consistent, the speech is meaningful, and the behavior is constructive.

The laws of quantum mechanics are not yet figured by us because of their changeable and etherical character, but they will be explained, no doubt, making *Albert Einstein's* thirty-year search of the Universal Structure of Life reality.

I wish I could Live then in the Unanswerable WHEN?

2. The Divine Spark of the Universal Consciousness!

Our new global culture demands now that we develop our social intelligence into the international one. We see and hear on mass media that people are often judged or even labeled from a narrow perspective of a dominant culture. They are manipulated with the help of money, a higher position, or just by having more energy physically.

We must be more culturally aware to decode each culture better to have a better idea of the international situation overall and the reasons that generate immigration. It must be essential in our schools to have **the News Briefings** when the students exchange the news from all over the world and their own country and ger better informed and, therefore, **better internationally connected**.

The chaotic and often politicized presentation of news on TV and the Internet generates ***an appalling unawareness*** of young people about the real state of things in their own country and in the world. There cannot be any level of intelligence formed in the obscurity of the state of things in science and in the world.

The world is our Common Family, and the state of things in the family is everyone's concern!

Lack of news connectedness and international awareness makes people believe whatever is presented to them, often in a fake framework. ***How can we develop creative, independent thinking in the mind of a young person who has no individual vision of the world*** in which he / she is living? No wonder, humanoids are so aggressive in their common intention to destroy humanity.

They are also under-informed about what evolved humanity is all about! Our country has the best scientists in the world, but our young people have no knowledge of the latest developments in science in a holistic way, not just in the subject which they study.

Don't teach just the subject. Teach the whole person! *(Leo Vygotsky)*

Disconnection adds fuel to the battle of our misconceptions and misunderstandings to the point that we are unable to integrate our different qualities, cultures, educations, values, religions, and politics. **This is the 21st century after all! Haven't we had centuries of ignorance to know better?** ***We evolve only through the process of integration and mutual enrichment****!* The lack of real **International Intelligence** hampers this process because we are not conceptually aware of what every culture constitutes in the Universal sense. We forget that the Earth has evolved to this point thanks to the contributions of its best minds from all over the world.

Our Evolutionary Human Refinery is Universal in its Digital Binary!

We need to honor not only our differences, but our commonness and sameness. *Our new transforming* ***Social Intelligence*** *necessitates breaking the vicious circle of behavioral standards, transcending our cultural and religious differences to develop a divine spark of* ***UNIVERSAL INTELLIGENCE*** *in us, based on acceptance and mutual respect for our common time and space on* ***Mother Earth.***

"Like Begets Like!"

3. Human Intelligence + Universal Intelligence = Super-Intelligence!

Next, to get our personal intelligence in synch with the Universal one, **we need to change the world of knowledge and the WORLD OF WILL** ourselves. As a great German philosopher, Emanuel Kant, puts it, ***"Will is the thing in itself, the inner content of knowing, the essence of the world."*** Will-power and self-discipline, in tandem, comprise are the main moving force on the road of getting free from the social programming of the **"COLLECTIVE UNCONSIOUS"** *(Carl Yung)*. Conscious use of willower paves the way to our ***feeling self-reliant, self-sufficient, and self-challenging*** and base these qualities on a much better awareness of the world and the latest developments in science

limited vision and generating a new- technologically-enhanced one that gives us a chance to tune our **AUTO-ANTENNAS** to the universal mind and receive the messages that it is emitting to us at every level of our consciousness development

To attain this stage of freedom, full of awareness and self-sufficiency, we need to do a lot of **WILLFUL SELF-SCULPTURING** ***that is the hardest job on earth!***

We need to master all vistas of intelligence at least at the dilettante level in sync with the will to resist the downfall tendencies that are tempting us all the time! A great German philosopher ***Schopenhauer*** writes,

"**In one aspect, the world is Universal entirely idea, in another one, it is entirely WILL**"

Intelligence exists in time and space; it is never static! It is in the constant state of flux that is changeable as the result of our changing thoughts, words, and actions. The more consciousness you bring into your life , ***the stronger your immune system*** is and the higher your spirit is

Body + Spirit+ Mind + Self-Consciousness + Universal Consciousness

The connecting element of our spiritual fractal is - **SPIRIT + FAITH** which are impossible to install into robot-humanoids. These are the basic glueing elements of our lives, making our priority absolute!

"A healthy spirit gives a healthy body, and a healthy body generates a healthy spirit! "So, learn to say "HALT!" to your whims and negative thought" *(Edgar Cayce)* Working at your will, you are moving from a three dimensional *(height, width, depth)* to **the five - dimensional consciousness** that demands that **we change the volume of our thinking** by willfully intellectualizing our emotions and expanding our consciousness to the universal level.

This evolutionary process requires a much higher level of our mental vibrations. Interestingly, we do not see the humanoids that are undoubtedly there somewhere because they exist on the higher overtone than we do.

"Let's Change the Fabric of Our Thinking!" *(Leo Vygotsky)*

4. Universal Luminosity is Our Velocity!

"God is a supercomputer that contains a total inventory of what each of us is."*(Rev. P. S. Berg)*
Finishing this book, I want to give tribute to many spiritually uplifted people from the inventory of God that had enlightened our lives with their inner light of wisdom and that continue to do so with their generous service on the path of enlightening us. They were / are, and will be not simply great, fabulous, fantastic, or even glamorous.

They are luminous. They are the Luminaries of our world!

There are many people carrying ***the godly spark of light*** in them around us. A skin color, nationality, a social position, or a person's legacy that might be even publicly acclaimed have nothing to do with it.

These people magnetize us with their unique wisdom and luminosity that outlines any common criticism or mass media incisures.

We see and hear a lot of Sadh guru lately, and I marvel at his individual wisdom and an immensity of his personality.

The inner light that such people radiate fills us up with their ***sincere faith, high values, and incredible tenacity*** .You cannot, but gravitate to such people in your mind and heart, and every encounter with a luminous person like that on the pages of his / her book, on TV, or in a face-to face communication, however short it may be, enlightens the memory for years to come.

Let there be many more luminaries without time and space boundaries!

The luminosity of ***Marilyn Monroe, Elvis Presley, or Michael Jackson***, for example, cannot be darkened by anybody ever. They have left to us the gift of inspiration and the impossibility to ever be duplicated in their uniqueness.

There is a well-known saying, "***Beauty will save the world!***" These words had been re-worded by a famous Russian painter, writer, theosophist, and enlightener, ***Nicholas Roerich***, whom I have quoted above.

"Beauty that is consciously perceived will save the world!"

Only when we can perceive the beauty of life around us and ***radiate it back in our thoughts, words, emotions, and actions*** will we be spiritually advanced to the Universal level of consciousness. We will integrate life in us in its entirety, and the legacy that we might leave behind will radiate the luminosity of our souls to the people that will carry this torch of light after us, like Prometheus.

That's our legacy! Everything is left for the people coming after us!

Auto-Induction:

I Am an Active Captain of My Life. I Am Alive!

5. Unite with Universal Intelligence Might!

To conclude the book, let me remind you to continue ***doing suggestive inspiring yourself*** at any time and space . Every time you are out in the sun or while enjoying the view of the night sky, see yourself, surrounded by light. ***Feel yourself immersed in a luminous substance*** - the Sun rays, the white light, and the light of the stars, even the light of every lamp you are looking at or passing by while driving .at night.

If you happen to be walking, ***breathe in light consciously at every step,*** saying any autosuggestion that you might happen to memorize. **Breathe in light, love, peace, and harmony**; ***breathe out negativity, troubles, sickness, problems, and darkness.***

Make such habit your second nature. Enjoy doing it!

After you have enlightened the consciousness of your entire body, you can start purifying it on the cellular level, heightening your aware attention to the body's needs and its complaints to you in the form of pain, fatigue, depression, sadness, lack of love, any insult, disappointment, inner emptiness, and loneliness.

Gradually, you will establish a true ***"mind-body connection"*** at your own accord. Practicing **Suggestive Meditation,** you will not wear your body out prematurely because you are inspiring it for life ***consciously, continuously, and religiously!***

Give your SELF-CONSCIOUSNESS a much better home to operate from!

Reading self- help books about how to preserve the vitality of the body, working on its mobility and beauty, we heighten the level of our ***general, physical intelligence that is fantastic as it is ,*** and we look further over the horizon of life, becoming conscious of the metaphysical realm that surrounds us.

Developing your physical intelligence consciously, we are also becoming increasingly aware of the magic of life in us.

Even if you do not feel that there is anything to thank your body for, keep your aware attention on it religiously. ***There is no need to better your overall intelligence*** *(10 stages, outlined above)* ***if you do not take care of your physical one properly.*** Thank your body for ***its unified improvement of*** your health every single day, good or bad, healthy, or sick, happy, or unhappy. **The body is doing it, anyway!**

Your Body is Your Temple!

Congratulations! You are the Living Intelligence in Action!

Conclusion to the System Infusion

Intelligence-
Ascending is
Self-Consciousness-
Mending!

Be a Thought-Evolved Being with the Universal Seeing!

Expand Your Life's Range - Change!

1. Holistic Education is the Solution for Our Evolution!

In summary, our educational goal is to gradually move to the universal realm of extra-terrestrial communication and become the citizens of the universal realm of the mind ***- the Source !***New amazing discoveries are made everyday thanks to the ***Artificial Intelligence*** that is propelling us ***up-ward, on-ward, and God-ward***.

The development of human consciousness is inseparable from **Digital – Quantum Evolution** that is pushing us forward in terms of ***creating a new volume of our consciousness*** *that* will expand our communication with each other and the world around us inconceivably.

No doubt, Artificial intelligence is channeling our evolution!

Modern times need constructive, well-structured, properly channeled, emotionally empowered minds that will necessitate ***physical, emotional, spiritual, and universal changes*** and propel us forward to the Star People level.

The fantastic thought-machine of yours is not for the residence of moths!

There is *a great old Jewish* saying that illustrates the idea of a conceptual mind-set that we need to generate,

"If there are only two courses of action, you should always take the third one."

In our case, it is self-development along the paths of **Holistic Education**, the education that provides **DEVELOPMENTAL INTELLIGENCE** in five essential life dimensions that will expand our human evolutionary outreach and raise our self-consciousness:

Physical + Emotional + Mental + Spiritual + Universal!

Unconscious ⟶ Conscious ⟶ Superconscious

Self-Awareness – Self-Monitoring - Self-Installation – Self-Realization - Self-Salvation!

Such an education, backed up by an exponential growth of technology and supplemented by an irresistible urge for ***self-installation in life, should follow the systemic formula:***

Self-Synthesis – Self-Analysis - Self-Synthesis!

Generalize Knowledge - Internalize it - Personalizing it – Strategize it -Actualize IT!

Our Digitally Enhanced Human Essence is in the Holistic Renaissance!

2. Intellectual Spiral of Self-Resurrection

Ten Stages of Living Intelligence, AI enhanced and Quantumly devised.

That's only the Starting Point!

Universal Intelligence completes ***the Self-Actualization Pyramid*** of the ten essential levels of ***Living Intelligence*** in the mental realm of the spiral of self-resurrection that I have outlined for you in ***in the books on Inspirational and Digital psychology for Self-Ecology***. All the stages are interconnected in the holistic system, presented stage by stage in five life realms- **physical** + **emotional** + **mental** + **spiritual** + **universal,** with two vistas of intelligence in each one.

10. Universal Intelligence — **Super-Level** of Consciousness / **Stage 5**
9. Spiritual Intelligence — *(Universal Dimension)*

8 Social Intelligence — **Macro- Level / Stage 4**
7. Cultural Intelligence — *(Spiritual Dimension)*

6. Financial Intelligence — **Mezzo-Level / Stage 3**
5. Professional / *** *Creative* ***/ Intelligence — *(Mental Dimension)*

4. Psychological Intelligence — **Meta- Level / Stage2**
3 Emotional Intelligence — *Emotional Dimension)*

`2. Language Intelligence — **Mini-Level / Stage 1**
1. General Intelligence *(Self-Genesis)* — *(Physical Dimension)*

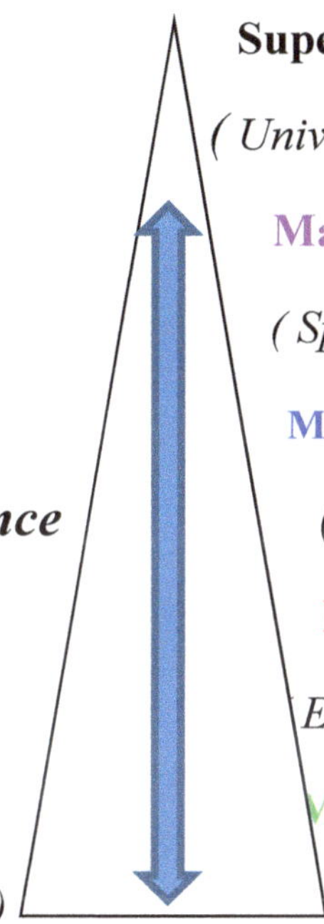

The Spiral of Your Self-Resurrection is never completed!

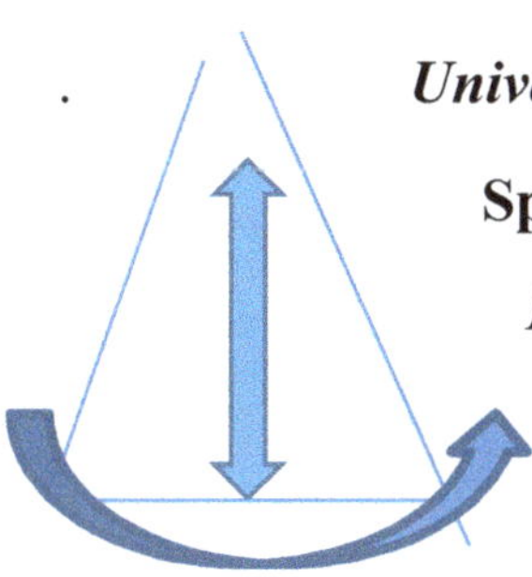

Universal level	**Super-Consciousness**	***Self-Salvation***
Spiritual level	**Self-Consciousness**	***Self-Realization***
Mental level	**Mind**	***Self-Installation***
Emotional level	**Spirit**	***Self-Monitoring***
Physical level	**Body**	***Self- Awareness***

Body+ Spirit+ Mind + Self-Consciousness+ Universal Consciousness =

Being Whole Makes Us Eternal!

Oh, Gee; I'm the Quality Me!

(Picture by Yolanta Lensky)

A New, Holistically Educated and Intellectually Spiritualized Cosmic Self!

<u>Post Thought</u>

"I Have a Dream!"

(Martin Luther King Jr.)

When we arrived in the USA in 1994, there were many things that we lacked in the former Soviet Union of those destructive times , but the country of plenty lacked one thing that my 12-year-old daughter noticed very soon and told me one day after she came home from Mamaroneck school. in Westchester.

-*"Mom, I know my dream now. I will start a business in this country;* ***I will build a center for kids!*** *They have no Pioneer Canters for kids, where I learnt folk-dancing, ballet-dancing, sewing, fishing, swimming, space-flying ,and toy-making.* ***Poor kids. They don't have that here. Nor do they have anything for teenagers. What should I do here? Where do I go after school?*** *My school peers talk only about parties, bars, sex ,and sleepovers."*

I was glad that she started thinking in the American way business-wise, but I was truly upset that she was right about the second part of her observation that very soon became a true nightmare for me. Reality changed her dream, and even though she became a computer designer and got eight books for kids published, she did not manage ***to bridge her dream and reality and become totally self-realized.*** She had never spoken about it again.

Years went by, but I have never forgotten that conversation and ***the spark in Yolanda's eyes*** because I saw the same spark in the eyes of many of my students whom I try to inspire to never betray their dreams and become what they aspire to be, not what just secures their life As a matter of fact, my five books on **Self-Resurrection** that this one completes are inspired by my daughter's unrealized dream to have the center like that someday here .

My dream is to have a free center " **JUPITER"** that will be structured in five rings - ***physical, emotional, mental, spiritual, and universal***, opening the doors to five make-believe **SPACE STATIONS** to inspire kids.

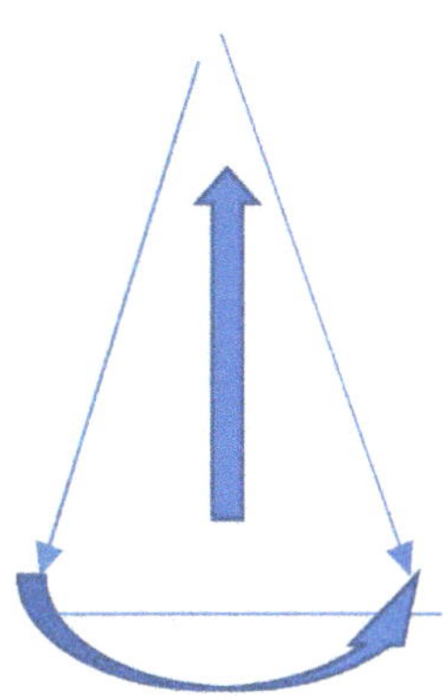

5. Universal Space Station

4. Spiritual Space Station

3. Mental Space Station

2. Emotional Space Station

1. Physical Space Station

The Choices We Make Dictate the Life We Live!

Nothing is Impossible if We Make Our Self-Resurrection Irreversible!

Wouldn't it be great to have the **PSYSICAL SPACE STATION** for the kids and teenagers where they will be developing their ***physical abilities***, like it was in the ancient Greek - ***the School for Spartans***. They will have a chance to enrich their ***cognitive skills*** and learn many interesting new things of their choice, as well as train their ***digital skills*** in the most creative ways. *(the book "**I am Free to Be the Best of Me!")***

Next, at the **EMOTIONAL STACE STATION,** the kids will learn about an array of beautiful human emotions and how ***to handle their Amygdala gland*** with the help of the best movies, tales, works of art, and beautiful classic music There will be instructors – ***psychologists- emotionalists*** *and* ***robot -friends*** to back them up if they happen to swoon emotionally. *(the book "**Soul-Refining**")*

The holistic pyramid of essential ***Vistas of Intelligence*** will be at their fingerprints when they chose to spiral up to the **MENTAL SPACE STATION** to fill up the gaps in the levels of intelligence that this book features. ***They will enhance their self-awareness by connecting their two brains and boosting their creativity*** and ascertaining their dream of what they want to become *(the book " **Living Intelligence or the Art of Becoming**!")*

The most inspiring knowledge of the human evolutionary ***spiritual growth*** will await them at the **SPIRITUAL SPACE STATION,** with *no preaching or religious indoctrination*, but ***insightful, interesting, science-based enlightening of the minds*** to shape their ***spiritualized intelligence***.*(the book " **Self- Taming!")***

The space-time journey into the impossible will be completed at the **UNIVERSAL SPACE STATION,** with breath-taking make-belief flight of imagination in the top observatory (Dr. Neil deGrasse *Tyson's a virtual space station)* that will take them on interstellar travels into space. *(the book " Beyond the Terrestrial!)*

And the best thing of all - there will be no obligations , no words: "***You are supposed to..."/ "You need to pay first!" That's not your station!" You need to first do this.*** The possibilities are limitless, as well as our incredible **NOW** that is, in fact, our wonderous future!

Let the Minds of our kids fly beyond the Boundaries of the Sky!

Let them Radiate the Inner Bliss of Everyday Happiness Release!

Somewhere and Someday are Here, today!

Dr. Ray with Her Inspirational Say!

*1. "**Emotional Diplomacy** or **Follow the Bliss of the Uncatchable Is!"/ Editorial** LEIRIS, New York, USA,2001*

***2."Four Dimensions of a Soul**" (Auto-Suggestive Psychology in Russian) / LEIRIS Publishing, New York, USA,2011*

***3."Americanize Your Language, Emotionalize Your Speech!" /** Nova Press, USA, 2011*

***4."It Too Shall Pass!" (**Inspirational Psychology for Self-Ecology) / Xlibris, 2012*

5."I am Strong in My Spirit!"** (Inspirational Boosters in Russian) / Xlibris, 201**5.

***6. "Language Intelligence or Universal English"** (Method of the Right Language Behavior), **Book One** /Xlibris, 2013 - Also, Stonewallpress,2019*

***7. "Language Intelligence or Universal English**" **(**Remedy Your Language Habits," **Book Two** /Xlibris, 2013 – Also, Stonewallpress,2019*

***8. "Language Intelligence or Universal English,"** (Remedy Your Speech Skills) **Book Three** /Xlibris, 2013- Also, Stonewallpress,2019*

***9. My Solar System** (Inspirational Psychology for Self-Ecology), Second Edition- Page Turner, Press and Media, 2019*

*10. **Language Intelligence** (three books in one) Stonewall Press, 2019*

Books on the Holistic Self-Resurrection:

***11."I Am Free to Be the Best of Me!"- (Physical Dimension) -** Toplinkpublishing.com. Sept. 2017) – Second Edition, Book Whip, 2019*

***12. Soul-Refining! (Emotional Dimension) (**Toplinkpublishing.com. May 2017) - Second Edition 2ook Whip, 2019*

***13. "Living Intelligence or the Art of Becoming**" (Mental **Dimension)-** (A New Paradigm of Self-Creation) Xlibris, 2015 - Second Edition , Bookwhip,2019*

***14. "Self-Taming" (**Life-Gaining is in Self-Taming! **/ Spiritual Dimension)-** Book Whip, 2019*

***15." Beyond the Terrestrial!"** (Be the Station for Self-Inspiration**!) - (Universal Dimension)** Xlibris, June 2016- Second Edition -, Book Whip, 2019 / - URLink Print and Media, 2019*

*16. **The State of Love from the Above)** Book Whip, 2019*

***17. "Love-Ecology"** (Love Ecology for Self-Psychology**)-** De. Rimaletta Ray Publishing*

***18. "Self-Worth"- Parchment ,** New Your,2020*

19. **"Self-Renaissance" / Workbook press,** *Las Vegas, 2020*

20" Soul-Symmetry!"/ *Clever Publication,* *Canada, 2021*

21. ".Dis-Entangle-ment!"/ Ivy Lit Press, *New York, 2022*

23."Digital Binary Human Refinery=Super-Human! " / *Brookside Press 2023, Canada*

24. "Exceptionality!" / Workbook Press , *Las Vegas / 2023*

25. "Trans-Human Acculturation"/ *Brookside Press/* ***2023,*** *Canada*

26.*"*Transcendent Un and AIs"/ *Brookside Press/* ***2024,*** *Canada*

27. " Light is Me. Light is My Philosophy!" */Brookside Press/* ***2024,*** *Canada*

28. "Love Ecology, *Second Edition (Love Psychology for Self-Ecology)/* DR. R*imaletta Ray Publishing/ 2024*

Www.holisticself-resurrection.com

Www. Language – fitness.com

email; - dr. rimaletta@gmail.com

Tel (202)212-2673

Pictures and designs are by ***Yolanta Lensky –***

Www.spontendormedia.com

(203) 212-2673

"There are two most important days in your life – the day you were born, and the day you realized <u>What For</u>."

(Mark Twain)

www.ingramcontent.com/pod-product-compliance
Lightning Source LLC
LaVergne TN
LVHW070851210926
848482LV00016B/915